*One-dimensional Stefan
Problems: an
Introduction*

π Pitman Monographs and
Surveys in Pure and Applied Mathematics   31

One-dimensional Stefan Problems: an Introduction

James M. Hill

University of Wollongong

Longman
Scientific &
Technical

Copublished in the United States with
John Wiley & Sons, Inc., New York

Longman Scientific & Technical
Longman Group UK Limited
Longman House, Burnt Mill, Harlow
Essex CM20 2JE, England
and Associated Companies throughout the world.

Copublished in the United States with
John Wiley & Sons, Inc., 605 Third Avenue, New York, NY 10158

First published 1987

AMS Subject Classifications: (main) 35K99,80A20
                            (subsidiary) 45G10,65N99

ISSN 0269–3666

British Library Cataloguing in Publication Data

Hill, J. M.
    One-dimensional Stefan problems: an
    introduction.—(Pitman monographs and
    surveys in pure and applied mathematics,
    ISSN 0269-3666; 31)
    1. Boundary value problems   2. Variational
    principles
    I. Title
    515'.62   QA379
ISBN 0-582-98823-3

Library of Congress Cataloging-in-Publication Data

Hill, James M., 1945–
    One-dimensional Stefan problems.

    (Pitman monographs and surveys in pure and applied
mathematics, ISSN 0269-3666; 31)
    Bibliography: p.
    Includes index.
    1. Heat—Conduction.   2. Boundary value problems.
I. Title.   II. Series.
QC321.H63   1987   536'.23   86-20054
ISBN 0-470-20388-9 (USA only)

Typeset in 10/12 Times
Printed and Bound in Great Britain
at The Bath Press, Avon

For Desley, Emily, Ruth and Thomas

Contents

Preface

Moving-boundary problems involving heat or diffusion phenomena occur in numerous important areas of science, engineering and industry. For example, freezing and thawing foods, production of ice, ice formation on pipe surfaces, solidification of steel and chemical reactions all involve either a moving freezing, moving melting or moving reaction front which is unknown. Mathematically, these problems require solving the diffusion or heat-conduction equation in an unknown region which has also to be determined as part of the problem. Over the past hundred years the subject has acquired an extensive literature in practically every scientific and engineering discipline, the most important of which are possibly, heat and mass transfer, chemical engineering, metallurgy, soil science and more traditional disciplines such as chemistry, physics, biology and mathematics. There are few exact solutions to moving-boundary problems, and for heat-diffusion problems the only physically relevant exact solutions occur when the position of the moving boundary (or boundaries) varies as the square root of time (that is, similarity solutions). The mathematical difficulties associated with solving further moving-boundary problems exactly are considerable and are of a fundamental character.

The nature of mathematical analysis is such that if a problem cannot be solved exactly then there is unlimited scope for approximate solutions. This is certainly the case for moving-boundary problems where the mathematical literature on the subject has, roughly speaking, developed in three main areas, approximate analytical methods, numerical techniques and qualitative results such as existence and uniqueness. In each of these three areas the literature is extensive and the task of doing justice to one of them is daunting enough, let alone all three. For practical scientists and engineers primarily interested in simple formulae for applications, a book dealing with the relative merits of some of the existing semi-analytical procedures would seem to be both worthwhile and necessary. However, if we set about such a task then it becomes apparent that many of the various approximate analytic methods are either very complicated to calculate but accurate, or simple to calculate but arbitrary with uncertain results. Thus, many of the approximate methods have no real rational basis other than personal preference.

In an introductory text to the subject it would seem inappropriate to include all known approximate techniques. The strategy I have adopted is to include only those methods which are compatible or relate to the general theme of the book. Those devices not directly bearing on the development presented are simply noted. In terms of approximate analytic solutions, the pseudo-steady-state approximation, large Stefan number expansion, integral formulation, upper and lower bounds for the motion, formal series solutions, unsolved integral equations for the motion and integral iteration techniques form not only a coherent and inter-connected development but represent concepts which will remain part of the subject irrespective of future developments. In addition, I have also included material on Goodman's and Megerlin's methods and on an iterative series solution procedure, which, although separate from the main theme of the book, provide important approximations which may be compared with the results of preceding chapters. In a sense the book represents a record of developments of the subject from this department together with those related results from the literature. As such it is appropriate to acknowledge the contributions of past and present students. Gregory B. Davis for initiating my interest in moving-boundary problems, Jeffrey N. Dewynne for his insight and willingness to provide much of the graphical and numerical material and, finally, to Adam Kucera for many helpful discussions.

This book is unusual in the sense that it deals exclusively with the classical single phase one-dimensional Stefan problems. These problems constitute the simplest non-trivial problems which display all the mathematical difficulties inherent in more complicated and perhaps more interesting heat-diffusion moving-boundary problems. The book assumes some knowledge of either heat conduction or diffusion and some familiarity with either Carslaw and Jaeger, *Conduction of Heat in Solids,* or Crank, *The Mathematics of Diffusion.* I have attempted to pitch the book at a fairly elementary level so that it may be suitable either for third- or fourth-year students in science, engineering or mathematics or for practical scientists and engineers who are not primarily mathematicians.

Chapter 1 presents a review of the physical problems and mathematical development associated with simple heat-diffusion moving-boundary problems. The main problems studied in the book are formulated in the second section of the chapter. The next two sections deal with the exact Neumann solution and the pseudo-steady-state approximation. This approximation is important primarily because of its simplicity but also because the pseudo-steady-state motion is subsequently established as a lower bound to the actual motion. The chapter also gives a review of those semi-analytic devices which are not included in any detail in subsequent chapters. Chapters 2, 3 and 4 develop the main theme of the book with reference to the classical one-dimensional Stefan problems. In Chapter 2 solutions for large Stefan numbers are presented. For the freezing of infinite circular cylinders or spheres, although the pseudo-steady-state approximation emerges as the leading term, subsequent corrections to the temperature profile are in fact singular at the point of complete freezing and therefore the results presented represent only partial solutions. However, the order one correction to the pseudo-steady motion is well defined up to and including complete freezing and this corrected motion turns out to be an upper bound to the actual motion and this is the link with Chapter 3.

In Chapter 3 derivations are given of the basic integral formulations for the classical one-dimensional Stefan problems, firstly, by direct integration and, secondly, using Green's functions. This integral formulation enables a formal integration of the equation for the motion of the boundary which although not explicit may be utilized to obtain upper and lower bounds to the motion. For example, simple and useful bounds follow immediately even from the physically obvious inequalities $0 \le T(r, t) \le 1$. The upper bound to the motion may be improved by showing that the actual temperature is bounded above by the pseudo-steady-state temperature. The lower bound to the motion is improved by a further integration of the basic equation for the boundary motion. The simple nature of these bounds and the known relation with the actual solution will ensure that the results and techniques of Chapter 3 are of permanent interest. Further consequences of the basic integral formulations of Chapter 3 are developed in Chapter 4. Firstly, formal series solutions are established which lead to unusual and unsolved integral equations for the motion of the moving boundary. Secondly, an integral iteration procedure is formulated and solutions are obtained employing the pseudo-steady-state approximation as an initial estimate. Thus Chapters 3 and 4 describe basic integral formulations and their consequences.

Chapter 5 is a departure from the development given in previous chapters and describes approximate methods based on assuming polynomial temperature profiles. Goodman's method is widely used in the literature and is approximate in the sense that the heat-diffusion equation is satisfied on average only. However, for the problems considered, Megerlin's method is the simpler. Results from these techniques are compared with other approximations. Chapter 6 describes a particular iterative series solution procedure for the classical one-dimensional Stefan problems. The method hinges on choosing appropriate boundary-fixing transformations which differ for the three different geometries under consideration. A large-time solution for the infinite slab with Newton cooling on the surface, and the solutions for freezing cylinders and spheres with no Newton cooling, give rise to particularly accurate numerical results. An alternative partial solution for the sphere is presented to give a possible coherent picture of the various boundary-fixing transformations. Chapter 7 is designed to demonstrate the utility of the ideas presented in previous chapters for a slightly more complicated problem. This chapter deals with moving-boundary problems involving both a fast and slow chemical reaction. The integral formulations obtained by simple integration and the Green's function approach are shown to yield distinct results in this case, the bounds resulting from the Green's function method being slightly superior. Some general formulae for Langford's functions $c_n(z, z_0)$ and $e_n(z, z_0)$, which arise in general solutions of the heat equation with cylindrical geometry, are given in Appendix 1 and some calculations relating to Chapter 6 are outlined in Appendix 2.

Whatever personal views exist as to the best method of solving moving-boundary problems, it remains a fact that students are attracted by the simplicity and relevance of the subject. Moreover, experience with the diffusion or heat-conduction equation readily translates to mathematical modelling in an employment situation. I would like to think that such courses in mathematical analysis, together with training in numerical analysis and modelling, form a sensible basis for aspiring mathematical modellers.

JAMES M. HILL
Department of Mathematics
The University of Wollongong

Acknowledgements

It is with much pleasure that I acknowledge Dr G. B. Davis and Dr V. G. Hart for carefully reading a preliminary version of this book and providing extensive lists of omissions, spelling mistakes and corrections.

I am also grateful to the *Quarterly of Applied Mathematics* and the Australian Mathematical Society for their kind permission to reproduce material from my publications.

List of symbols

The following symbols have the same meaning throughout the book and the equation given is where first used. Additional symbols used are listed at the end of each chapter.

a	convenient length scale, cylindrical or spherical radius of container
C_1	specific heat of solid, (1.3)
C_2	specific heat of liquid, (1.5)
E	emissivity of surface, (1.10)
h	surface heat transfer coefficient, (1.9)
k_1	thermal conductivity of solid, (1.3)
k_2	thermal conductivity of liquid, (1.5)
L	latent heat of fusion, (1.8)
r	non-dimensional cylindrical or spherical position radius, (1.28)
r^*	physical cylindrical or spherical position radius, (1.25)
$R(t)$	non-dimensional cylindrical or spherical moving boundary, (1.28)
$R^*(t^*)$	physical cylindrical or spherical moving boundary, (1.25)
t	non-dimensional time, (1.18)
t^*	physical time, (1.1)
t_c	time to complete solidification of sphere and cylinder
$t_{pss}(R)$	pseudo-steady-state motion, (1.67)
t_{pssc}	pseudo-steady-state estimate of t_c, (1.79)
T_f	uniform freezing or fusion temperature, (1.1)
T_ℓ	uniform initial liquid temperature, (1.1)
T_0	uniform surface temperature, (1.1)
$T(x, t)$	non-dimensional solid temperature for slab, (1.19)
$T(r, t)$	non-dimensional solid temperature for cylinder or sphere, (1.29)

$T_1^*(x^*, t^*)$ physical solid temperature for slab, (1.3)
$T_2^*(x^*, t^*)$ physical liquid temperature for slab, (1.5)
$T_1^*(r^*, t^*)$ physical solid temperature for cylinder and sphere, (1.25)
$T_{pss}(x, t)$ pseudo-steady-state temperature for slab, (1.66)
$T_{pss}(r, t)$ pseudo-steady-state temperature for cylinder and sphere, (1.70)
$u(\rho, \tau)$ temperature T (or rT for sphere), (2.63), (2.79) and (2.94)
x non-dimensional position, (1.18)
x^* physical position, (1.1)
$X(t)$ non-dimensional boundary position, (1.18)
$X^*(t^*)$ physical boundary position, (1.2)

Greek symbols

α Stefan number or phase change parameter, (1.20)
β inverse of Biot modulus (i.e. B_i^{-1}), (1.20)
γ positive root of transcendental equation (1.43), (1.39)
κ_1 thermal diffusivity $k_1/\rho_1 C_1$, (1.11)
λ 0, 1 and 2 for slab, cylinder and sphere, respectively, (1.34)
ξ integration variable
ρ boundary fixing transformation, (2.8), (2.9) and (2.10)
ρ_1 density of both solid and liquid phases, (1.3)
σ_1 Stefan–Boltzmann constant, (1.10)
τ boundary position time variable, (2.63), (2.79) and (2.94)

Conventions

Stars or asterisks (*) are used to denote physical quantities. $T^\dagger(r, R)$ denotes the temperature $T(r, t)$ but employing the boundary position $R(t)$ as the time variable.

1

Heat-diffusion moving-boundary problems

1.1 Introduction

In 1889 J. Stefan published altogether four papers on problems involving an unknown moving boundary. The first and third papers involve heat conduction and changes of phase while the second and fourth describe, respectively, diffusional transport of material in a reaction zone and evaporation or condensation. The titles of the four papers are as follows:

(a) On some problems in the theory of heat conduction.
(b) On the diffusion of acid and alkaline solutions through each other.
(c) On the theory of ice formation with reference to the Arctic Sea.
(d) On evaporation and condensation as diffusion processes.

The first paper, Stefan (1889a), deals with the freezing of ground and the following two problems are posed and solved:

(i) The semi-infinite half-space $(0, \infty)$ consists of a material which may exist in either a liquid or solid phase. Initially the material is in the liquid phase at a known uniform temperature. At time $t^* = 0$ the surface $x^* = 0$ is cooled and subsequently maintained at a given constant temperature below that of the freezing temperature of the liquid. After time t^* the problem is to determine the thickness of the solid layer and the temperature in both the solid and liquid phases.

(ii) The infinite space $(-\infty, \infty)$ again is assumed to consist of a material which may exist in either a liquid or solid phase. Initially it is assumed that the region $(-\infty, 0)$ is in the solid phase and maintained at some known constant temperature below the freezing temperature, while the region $(0, \infty)$ is assumed to be in the liquid phase and also with a known uniform temperature throughout. Again, after time t^* the problem is to determine the position of the boundary separating the solid and liquid phases and the temperatures in both phases.

Assuming constant thermal properties of both the solid and liquid phases and negligible volume changes in freezing or melting, Stefan obtains exact analytical solutions to both problems. Both problems admit similarity solutions such that the position of the moving boundary varies as the square root of time and the constant of proportionality is determined from a certain transcendental equation.

As a result of these papers and another with exactly the title of (c) (Stefan 1891), problems involving changes of phase and moving surfaces of separation between phases are now loosely classified as 'Stefan problems'. This is despite the fact that Lamé and Clapeyron (1831) actually were the first to study the problem of determining the thickness of the solid crust generated by the cooling of a liquid under a constant surface temperature. Indeed, these authors identify the square root time thickness and give the single-phase exact solution (1.42) and (1.43) but, as pointed out by Brillouin (1931), do not actually give numerical values of the roots γ of (1.43). Apart from this, Lamé and Clapeyron (1831) were certainly the first authors to present this exact similarity solution. However, nowadays, this solution and the more general solutions utilized by Stefan are usually credited to Franz Neumann, as given in unpublished lectures delivered in Königsberg in the early 1860s (see Weber 1919). Neumann's solutions are remarkable for two reasons. Firstly, despite nearly one hundred years of mathematical research on Stefan problems, they and various extensions remain the only physically interesting exact solutions. A limited number of other exact solutions do exist but not to problems of any genuine practical interest. Secondly, it is remarkable the extent to which essentially Neumann's similarity solutions have been utilized within the context of different physical phenomena, perhaps involving multi-phases or multi-component mixtures and even accommodating changes in material properties such as diffusivities and densities. Common to all these varied physical situations is a slab geometry, infinite in extent with phase-change boundaries moving according to the similarity square root time law. For other geometries, such as finite slabs, cylinders and spheres, there are as yet no exact closed-form solutions to Stefan problems of physical interest.

In all areas of science and engineering an extensive literature on heat-diffusion moving-boundary problems has developed. Mathematical contributions to the subject have progressed in three main areas, approximate analytical methods, numerical techniques and qualitative results such as existence and uniqueness. In this book we attempt to give some indication of developments in the first area for a restricted class of idealized moving-boundary problems, which may be termed the classical one-dimensional Stefan problems. The problems are idealized in a number of ways, such as assuming constant physical properties and negligible volume changes on freezing or melting, but principally we assume through-out subsequent chapters that initially the material has acquired its uniform fusion or melting temperature. This means that instead of having two heat conduction or diffusion problems with a moving surface separating them, we have, with this assumption, only one such problem to solve. Of course, in reality there are very few problems for which all these various assumptions are physically meaningful. However, from our point of view such assumptions enable us to focus on the essential mathematical difficulties inherent in moving-boundary problems. More realistic problems perhaps involve many phases, multi-component mixtures, several moving boundaries and variable physical properties, such as densities and thermal conductivities. In order to make progress with problems of this complexity we need first to develop our intuition by means of the simplest non-trivial problems in the subject.

The only definitive text on free- and moving-boundary problems is the recent book by Crank (1984). This book presents a broad account of the subject and includes formulations of a large number of free- and moving-boundary problems, existing analytical approximations and numerical techniques appropriate to such problems. It contains extensive detailed information and references related to these areas and provides the only global perspective of mathematical developments in the subject of free- and moving-boundary problems. Although Crank (1984) is by far the most important text on moving-boundary problems there are several other books the reader may consult. Flemings (1974) treats the fundamentals of solidification processing from a physical and practical viewpoint. The book provides an excellent account of the basic scientific principles underlying heat flow, mass transport and interface kinetics leading to the solidification process. Carslaw and Jaeger (1965) and Crank (1964) each contain an informative chapter on moving-boundary problems. Carslaw and Jaeger (1965) discuss the subject from the point of view of heat flow and changes of state and present a compact account of the main analytic results available. Most of the chapter details Neumann's solution and its generalizations but it also includes other special solutions and numerous important references. Crank (1964), on the other hand, develops the known results and literature relating to diffusion with a moving boundary. In addition to the progressive freezing of a liquid, Crank (1964) considers the following three examples of diffusion in two distinct regions separated by a moving boundary or interface:

(i) The absorption by a liquid of a single component from a mixture of gases.
(ii) Tarnishing reactions in which a film of tarnish is formed at the

surface of a metal by reaction with a gas, the diffusion of gas through the film being the rate-controlling process.
(iii) Diffusion with a diffusion coefficient which is a discontinuous function of the concentration.

These problems are solved using Danckwerts (1950) general technique. In the first two examples there is a jump in the concentration across the moving boundary, while for the progressive freezing of a liquid and the third example there is a jump in the gradient of the concentration across the boundary. Elliott and Ockendon (1982) is essentially a book on numerical methods and Rubinstein (1971) is essentially concerned with qualitative results. However, both books give extensive descriptions of the underlying physical problems. In fact, the first two chapters of Elliott and Ockendon (1982) and the first four chapters of Rubinstein (1971) provide extremely useful background reading to the subject. In addition, each of these books contains over two hundred references, many of which are not reproduced here. Ockendon and Hodgkins (1975) and Wilson *et al.* (1978) are both proceedings of conferences on moving-boundary problems and are useful to obtain some insight into the range and variety of current activity in the area. From the point of view of this book, Ockendon and Hodgkins (1975) contains two particularly relevant articles, Tayler (1975) on the mathematical formulation of Stefan problems and Ockendon (1975) on various techniques of analysis for moving-boundary problems. Two recent books relating to the numerical treatment of moving-boundary problems are Lewis and Morgan (1979) and Albrecht *et al.* (1982).

There are, fortunately, several excellent review articles on the subject of moving-boundary problems. Bankoff (1964), Muehlbauer and Sunderland (1965) and Goodman (1964) are all particularly well written and informative. The first two deal exclusively with heat conduction or diffusion and changes of phase, while the latter surveys the application of integral methods to transient non-linear heat transfer, including problems involving a change of phase. All three contain extensive lists of references. More recent surveys of the subject are Furzeland (1977) and Rubinstein (1979). Furzeland (1977) contains considerable information on numerical techniques for both moving- and free-boundary problems. Rubinstein (1979) reviews the subject, emphasizing many of the still unsolved aspects of Stefan problems. Surveys of the subject are also given by Boley (1972), Cohen (1971) and Mori and Araki (1976).

There are numerous non-mathematical articles relating to specific moving-boundary problems which arise from technically important areas. The following articles represent only a small fraction of the literature on such problems and are included for purposes of illustration. A better appreciation of existing literature can be gauged from the references in either Elliott and Ockendon (1982) or Rubinstein (1971) or from the review articles by Bankoff (1964) and Muehlbauer and Sunderland (1965). Chalmers (1954) describes the basic metallurgical viewpoint of melting and freezing with particular reference to the structure of both solidified pure metals and alloys. Field (1927) gives a simple formula for the case of a semi-infinite mould in contact with a semi-infinite liquid, assuming steel solidifies at a single temperature, that the thermal properties of the mould and casting are identical and that contact between the mould and casting is maintained throughout the solidification process. Paschkis (1953) provides valuable experimental and theoretical data on the solidification of both long and short steel ingots, with a view to estimating the cooling effects of the ends of the cylinder on the rate of solidification. Dewey et al. (1960) describe an implicit finite difference technique for the transient heat conduction in a long cylinder, but of finite thickness with a non-linear thermal conductivity and a moving boundary. A specific application is made for the problem of re-entry heating encountered by aerodynamic bodies for which erosion occurs due to ablation and combustion at the surface of the projectile. Elmer (1932), Pekeris and Slichter (1939) and Hwang (1977) all consider problems associated with ice formation on pipe surfaces. Elmer (1932) obtains the pseudo-steady-state solution of the problem which he relates to practical situations. Pekeris and Slichter (1939) obtain a first-order correction to the pseudo-steady-state solution, which is evaluated and found to be small. The solution obtained is applied to the problem of freezing soil around a long pipe. Hwang (1977) examines the validity of the two-dimensional solution for a buried pipe given by Carslaw and Jaeger (1965) with reference to numerical techniques. A modification of the solution is suggested for engineering purposes. Dana and Wheelock (1974) describe interesting experiments in which the acid elution of dark blue cupric amine complex from a cation-exchange resin produce sharply defined moving boundaries within each resin bead. The acid elution of blue copper amine from a strong-acid resin bead proceeds by a shrinking core process, which can be readily followed because of a distinct colour change. A theoretical model is shown to apply at small acid concentrations only. Zener (1949) gives the similarity solution involving the square root of time for the radius of a spherical precipitate particle growing in a solid solution of initially uniform composition. This solution is the spherical analogue of Neumann's solution. Tayler (1982) is an excellent general introduction to phase-change problems. Specific problems discussed are continuous casting, the shape of meltpools, resistance spot welding, vaporization of a liquid in contact with a heated solid block and condensation of a binary mixture on a cooled surface. In addition, a speculative model for the solidification of a binary alloy is discussed in detail, and phenomena associated with non-linear diffusion coefficients and diffusion coefficients which change sign are also discussed.

The above articles indicate the scope and applicability of moving-boundary analysis. In the following section of this chapter we formulate the classical one-dimensional Stefan problems which form the main study of this book. We also introduce appropriate non-dimensional variables which we employ throughout subsequent chapters. In Section 1.3 we present in some detail Neumann's exact solution applicable to the ideal-ized single-phase problem. In Section 1.4 we derive the pseudo-steady-state approximations to the three problems formulated in Section 1.2. The exact Neumann solution is compared with the appropriate pseudo-steady-state estimate. In order to provide a balanced view of the methods described in this book we give, in Section 1.5, a brief review of other semi-analytic techniques and we close with a brief summary, in Section 1.6, of the main qualitative features of temperature profiles and boundary motions obtained from a numerical enthalpy scheme.

1.2 Formulation of classical Stefan problems

Consider Stefan's original problem of the semi-infinite half-space $(0, \infty)$ assumed to consist of a material which may exist in either a liquid or solid phase. Suppose the material is initially in the liquid phase at constant temperature T_ℓ, which is above the freezing or fusion temperature T_f. Assume that at time $t^* = 0$ the plane surface $x^* = 0$ is instantaneously cooled and subsequently maintained at a constant temperature T_0, which is below the freezing temperature T_f. Immediately, the liquid along $x^* = 0$ freezes and, subsequently, a freezing front moves progressively through the liquid such that behind the front the material is in the solid phase, while ahead of the front the material remains in the liquid phase. Suppose that the temperature of the solid at position x^* and time t^* is $T_1^*(x^*, t^*)$ and similarly let $T_2^*(x^*, t^*)$ denote the temperature of the liquid. Clearly, the following inequalities hold;

$$T_0 \le T_1^*(x^*, t^*) \le T_f \le T_2^*(x^*, t^*) \le T_\ell. \tag{1.1}$$

In addition, we suppose the position of the front at time t^* is given by

$$x^* = X^*(t^*), \tag{1.2}$$

where $X^*(0) = 0$.

Assuming that the physical and thermal properties of both solid and liquid phases remain constant throughout, that there is no volume change in freezing and that heat transfer takes place by conduction only, then may we formulate the following problems for the solid and liquid phases, respectively:

$$\frac{\partial T_1^*}{\partial t^*} = \frac{k_1}{\rho_1 C_1} \frac{\partial^2 T_1^*}{\partial x^{*2}}, \qquad 0 < x^* < X^*(t^*), \tag{1.3}$$

$$T_1^*(0, t^*) = T_0, \tag{1.4}$$

and

$$\frac{\partial T_2^*}{\partial t^*} = \frac{k_2}{\rho_1 C_2} \frac{\partial^2 T_2^*}{\partial x^{*2}}, \qquad X^*(t^*) < x^* < \infty, \tag{1.5}$$

$$T_2^*(\infty, t^*) = T_\ell, \tag{1.6}$$

where ρ_1 is the density of both solid and liquid phases, k_1 and C_1 are the thermal conductivity and specific heat of the solid, respectively, and similarly k_2 and C_2 are the thermal conductivity and specific heat of the liquid, respectively. In addition, both temperatures at the front are the freezing temperature T_f, and the front moves in such a way that its velocity is proportional to the jump in heat flux across the front. In other words, if L is the latent heat of fusion then when the front moves a distance dX^*, a quantity of heat $\rho_1 L\, dX^*$ is liberated and removed by conduction. Thus, altogether on the front we have

$$T_1^*(X^*(t^*), t^*) = T_2^*(X^*(t^*), t^*) = T_f, \tag{1.7}$$

$$k_1 \frac{\partial T_1^*}{\partial x^*}(X^*(t^*), t^*) - k_2 \frac{\partial T_2^*}{\partial x^*}(X^*(t^*), t^*) = \rho_1 L \frac{dX^*}{dt^*}, \tag{1.8}$$

and the latter equation is sometimes referred to as the Stefan condition. Equations (1.3)–(1.8) constitute the problem posed and solved by Stefan (1889a). The complete solution is discussed in detail by Carslaw and Jaeger (1965, see p. 285). A special case of this solution which relates to subsequent chapters is presented in the following section.

If, instead of prescribing the temperature on the surface $x^* = 0$, we wish to allow heat loss by either forced convection (Newton cooling) or by black-body radiation, then the surface condition (1.4) is replaced by a condition involving the surface heat flux. In the case of Newton cooling on the surface we have in place of (1.4) (see Carslaw and Jaeger 1965, p. 18)

$$k_1 \frac{\partial T_1^*}{\partial x^*}(0, t^*) = h[T_1^*(0, t^*) - T_0], \tag{1.9}$$

where h is referred to as the surface heat-transfer coefficient. In the case of black-body radiation on the surface we have (see Carslaw and Jaeger

1965, p. 21)

$$k_1 \frac{\partial T_1^*}{\partial x^*}(0, t^*) = \sigma_1 E[T_1^*(0, t^*)^4 - T_0^4], \tag{1.10}$$

where σ_1 is the Stefan–Boltzmann constant and E is the emissivity of the surface, which is defined to be the ratio of the heat emitted by the surface to that emitted by a black-body equivalent surface at the same temperature. If heat is lost by both forced convection and black-body radiation then the surface heat flux involves a linear combination of both terms in (1.9) and (1.10). This would be the case, for example, for problems dealing with ice formation on the surface of lakes. Problems involving a surface condition of the form (1.9) or (1.10) are considerably more complicated than that with surface condition (1.4). In the case when the heat flux is prescribed, a finite time must elapse before the surface acquires the freezing temperature T_f. This finite time is determined by first solving a standard heat-conduction problem for the liquid phase.

For example, in the case of Newton cooling at the surface, the temperature in the liquid up to the time when the surface acquired the freezing temperature T_f is well known (see Carslaw and Jaeger 1965, p. 70) to be given by

$$\frac{T_1^*(x^*, t^*) - T_0}{T_\ell - T_0} = \mathrm{erf}\left(\frac{x^*}{2(\kappa_1 t^*)^{1/2}}\right) + \exp\left(\frac{hx^*}{k_1} + \frac{h^2}{k_1^2}\kappa_1 t^*\right)$$
$$\times \mathrm{erfc}\left(\frac{x^*}{2(\kappa_1 t^*)^{1/2}} + \frac{h}{k_1}(\kappa_1 t^*)^{1/2}\right), \tag{1.11}$$

where $\kappa_1 = k_1/(\rho_1 C_1)$ is the thermal diffusivity and, as usual, erf and erfc denote the error and complementary error functions defined by

$$\mathrm{erf}(x) = \frac{2}{\pi^{1/2}} \int_0^x e^{-\xi^2}\, \mathrm{d}\xi, \tag{1.12}$$

$$\mathrm{erfc}(x) = 1 - \mathrm{erf}(x) = \frac{2}{\pi^{1/2}} \int_x^\infty e^{-\xi^2}\, \mathrm{d}\xi. \tag{1.13}$$

Thus the time t_i^* taken for $x^* = 0$ to acquire the freezing temperature T_f is determined as the positive root of

$$\frac{T_f - T_0}{T_\ell - T_0} = \exp\left(\frac{h^2}{k_1^2}\kappa_1 t_i^*\right)\mathrm{erfc}\left(\frac{h}{k_1}(\kappa_1 t_i^*)^{1/2}\right), \tag{1.14}$$

and the temperature in the liquid at this instant is obtained from (1.11) with t^* equal to t_i^*. This temperature profile now becomes an 'initial' condition for the ensuing freezing problem and introduces an additional complexity into the problem.

In order to focus on the essential moving-boundary aspects of the problem, we avoid this difficulty by simply assuming that, initially, the liquid is on the point of freezing, that is the temperature of the liquid is everywhere equal to the constant fusion temperature T_f. This means that whenever the surface temperature is lowered, freezing immediately takes place and the problem reduces to one of heat conduction in the solid phase alone. This important assumption, which in many cases can be physically realized, reduces a two-phase Stefan problem to a one-phase problem. Thus, with $T_l \equiv T_f$ the Stefan problem (1.3) and (1.5)–(1.8) with Newton cooling on the surface becomes altogether,

$$\frac{\partial T_1^*}{\partial t^*} = \frac{k_1}{\rho_1 C_1} \frac{\partial^2 T_1^*}{\partial x^{*2}}, \qquad 0 < x^* < X^*(t^*), \tag{1.15}$$

$$\frac{\partial T_1^*}{\partial x^*}(0, t^*) = \frac{h}{k_1}[T_1^*(0, t^*) - T_0], \qquad T_1^*(X^*(t^*), t^*) = T_f, \tag{1.16}$$

$$\frac{\partial T_1^*}{\partial x^*}(X^*(t^*), t^*) = \frac{\rho_1 L}{k_1} \frac{dX^*}{dt^*}, \qquad X^*(0) = 0. \tag{1.17}$$

For this problem we now introduce non-dimensional variables and constants α and β, which are defined in terms of physical parameters by

$$x = \frac{x^*}{a}, \qquad X = \frac{X^*}{a}, \qquad t = \frac{k_1 t^*}{\rho_1 C_1 a^2}, \tag{1.18}$$

$$T(x, t) = \frac{T_f - T_1^*(x^*, t^*)}{T_f - T_0}, \tag{1.19}$$

$$\alpha = \frac{L}{C_1(T_f - T_0)}, \qquad \beta = \frac{k_1}{ha}, \tag{1.20}$$

where a denotes any convenient length scale which might, for example, be the boundary position at certain time. The constant α is referred to as the Stefan number or the phase change parameter and is the ratio of the latent heat to the heat capacitance of the solidified phase. The constant β^{-1} is sometimes known as the Biot modulus and denoted by Bi. For metals initially at their melting points, Table 1.1 gives numerical values of α for various metals ideally assumed to cool in surroundings at 20°C. Clearly, for metals in this situation the α values are small. These values should be contrasted with a value of α of about 8.0 for water initially at 0°C and the surface suddenly cooled to about −20°C. Values of the constant β are more difficult to obtain since this involves the surface heat-transfer coefficient h which is critically dependent on the surrounding environment.

Table 1.1 α values of various metals ($T_0 = 20°C$)

	L (calories/ gram)	C_1 (calories/ gram °C)	T_t (°C)	α $L/C_1(T_t - T_0)$
Aluminium	92.4	0.2096	660.0	0.689
Antimony	24.3	0.0508	630.5	0.635
Bismuth	13.0	0.0304	269.0	1.717
Cadmium	14.0	0.0547	320.9	0.851
Cobalt	58.0	0.1030	1490.0	0.383
Copper	43.0	0.0909	1083.0	0.445
Gold	15.9	0.0303	1063.0	0.503
Lead	5.0	0.0302	327.4	0.539
Platinum	27.0	0.0324	1773.0	0.475
Silver	22.0	0.0560	960.5	0.418
Tin	14.6	0.0536	231.9	1.286
Zinc	26.6	0.0918	419.5	0.725

We remark that we have non-dimensionalized the temperature in the frozen layer such that $T(x, t)$ satisfies the inequalities

$$0 \le T(x, t) \le 1. \tag{1.21}$$

For freezing problems this choice may appear somewhat counter-intuitive. However, in the context of the equivalent diffusion problem, normalization of the concentration is an obvious simplification. From (1.18)–(1.20) we find that the moving-boundary problem becomes

$$\frac{\partial T}{\partial t} = \frac{\partial^2 T}{\partial x^2}, \qquad 0 < x < X(t), \tag{1.22}$$

$$T(0, t) - \beta \frac{\partial T}{\partial x}(0, t) = 1, \qquad T(X(t), t) = 0, \tag{1.23}$$

$$\frac{\partial T}{\partial x}(X(t), t) = -\alpha \frac{dX}{dt}, \qquad X(0) = 0. \tag{1.24}$$

In the diffusion context this problem describes that of a prescribed constant concentration with some leakage at the surface $x^* = 0$ and progressing by diffusion into a region of zero concentration. We observe for this problem that the constant β can be completely removed simply by using new variables x/β, X/β and t/β^2. This is due to the lack of a characteristic length for the problem. However, the above formulation is preferable since it is useful to view the exact Neumann solution, given in the following section, as applicable to the problem (1.22)–(1.24) in the limit of β tending to zero. It is also useful to maintain a correspondence with this problem and the cylinder and sphere problems for which the constant β cannot be scaled from the problem. The above problem (1.22)–(1.24), for which there is no exact solution for β non-zero, is one of the basic problems examined in detail throughout the remainder of the book.

We now consider similar problems of freezing liquids in cylindrical and spherical containers of radius a and also initially held at the uniform fusion temperature T_f. The diffusion analogue of the spherical problem is particularly important since many industrial processes involve the flow of a fluid over a packed bed of chemically reacting particles, which are usually assumed spherical. The resulting diffusion moving-boundary problem in each spherical particle is consequently a fundamental problem. The problem of freezing a liquid at its fusion temperature T_f in a spherical container by means of a coolant at temperature T_0, and allowing Newton heat loss at the surface, becomes

$$\frac{\partial T_1^*}{\partial t^*} = \frac{k_1}{\rho_1 C_1}\left(\frac{\partial^2 T_1^*}{\partial r^{*2}} + \frac{2}{r^*}\frac{\partial T_1^*}{\partial r^*}\right), \qquad R^*(t^*) < r^* < a, \tag{1.25}$$

$$-k_1\frac{\partial T_1^*}{\partial r^*}(a, t^*) = h[T_1^*(a, t) - T_0], \qquad T_1^*(R^*(t^*), t^*) = T_f, \tag{1.26}$$

$$k_1\frac{\partial T_1^*}{\partial r^*}(R^*(t^*), t^*) = \rho_1 L\frac{dR^*}{dt^*}, \qquad R^*(0) = a, \tag{1.27}$$

where $T_1^*(r^*, t^*)$ denotes the temperature in the solid layer at position r^* and time t^*, and $R^*(t^*)$ denotes the radius of the freezing front at time t^*. On introducing non-dimensional variables

$$r = \frac{r^*}{a}, \qquad R = \frac{R^*}{a}, \qquad t = \frac{k_1 t^*}{\rho_1 C_1 a^2}, \tag{1.28}$$

$$T(r, t) = \frac{T_f - T_1^*(r^*, t^*)}{T_f - T_0}, \tag{1.29}$$

the temperature is again normalized such that (1.21) holds and the problem (1.25)–(1.27) becomes

$$\frac{\partial T}{\partial t} = \frac{\partial^2 T}{\partial r^2} + \frac{2}{r}\frac{\partial T}{\partial r}, \qquad R(t) < r < 1, \tag{1.30}$$

$$T(1, t) + \beta\frac{\partial T}{\partial r}(1, t) = 1, \qquad T(R(t), t) = 0, \tag{1.31}$$

$$\frac{\partial T}{\partial r}(R(t), t) = -\alpha\frac{dR}{dt}, \qquad R(0) = 1, \tag{1.32}$$

where the constants α and β are again given by (1.20). In a completely analogous manner the moving-boundary problem for freezing a liquid in an infinite circular cylindrical container, initially at its fusion temperature T_f, becomes

$$\frac{\partial T}{\partial t} = \frac{\partial^2 T}{\partial r^2} + \frac{1}{r}\frac{\partial T}{\partial r}, \qquad R(t) < r < 1, \tag{1.33}$$

together with (1.31) and (1.32).

These problems constitute the basic one-dimensional Stefan problems for cylinders and spheres for which no exact solutions are available. Of principle interest for these problems is the determination of the motion of the boundary $R(t)$ and, in particular, the time t_c for complete freezing, that is $R(t_c) = 0$. The constants α and β are strictly positive and non-negative, respectively, and roughly speaking increasing α and β both correspond to increased solidification or freezing times. Although it is usually necessary to treat these problems separately, it is sometimes useful to characterize all three by the single governing equation

$$\frac{\partial T}{\partial t} = \frac{\partial^2 T}{\partial r^2} + \frac{\lambda}{r}\frac{\partial T}{\partial r}, \qquad R(t) < r < 1, \tag{1.34}$$

together with (1.31) and (1.32), where $\lambda = 0$, 1 and 2 for the half-space $(-\infty, 1)$, infinite circular cylinder and sphere, respectively, and, for λ zero, x and $X(t)$ in (1.22)–(1.24) relate to r and $R(t)$ by the equations

$$r = 1 - x, \qquad R(t) = 1 - X(t). \tag{1.35}$$

There are of course numerous other important moving-boundary problems which also require detailed study. However, the above three problems highlight, in simple terms, the inherent difficulties common to all heat-diffusion moving-boundary problems. Moreover, a good deal of the mathematical literature on Stefan problems has been developed for these particular problems. However, we cannot inform the reader that all the developments associated with these particular Stefan problems can be readily extended to other heat-diffusion moving-boundary problems. As with progress in other non-linear areas, most of the advances made are frequently *ad hoc* and not readily generalized. However, it is likely that at least some of the ideas and techniques developed will apply to other problems.

1.3 Neumann's exact solution

In this section we give Neumann's exact solution to the moving-boundary problem (1.22)–(1.24) with no Newton's heat loss at the surface $x = 0$, that is the constant β is zero. We give two derivations. Firstly, we observe (1.22)–(1.24) with β zero remains invariant under the one-parameter group of transformations

$$\bar{x} = e^\varepsilon x, \qquad \bar{X} = e^\varepsilon X, \qquad \bar{t} = e^{2\varepsilon}t, \qquad \bar{T} = T, \tag{1.36}$$

where ε denotes the arbitrary parameter (see, for example, Hill 1982). Thus, if the solution takes the form

$$T = \Phi(x, t), \qquad X = \Psi(t), \tag{1.37}$$

then

$$\bar{T} = \Phi(\bar{x}, \bar{t}), \qquad \bar{X} = \Psi(\bar{t}), \tag{1.38}$$

and from (1.36) we can deduce in the usual way the functional form

$$T = \phi\left(\frac{x}{t^{1/2}}\right), \qquad X = (2\gamma t)^{1/2}, \tag{1.39}$$

where γ is a constant. With $\zeta = x/t^{1/2}$ the problem (1.22)–(1.24) for β zero becomes

$$\phi''(\zeta) + \frac{\zeta\phi'(\zeta)}{2} = 0, \tag{1.40}$$

$$\phi(0) = 1, \qquad \phi((2\gamma)^{1/2}) = 0, \qquad \phi'((2\gamma)^{1/2}) = -\alpha\left(\frac{\gamma}{2}\right)^{1/2}, \tag{1.41}$$

where primes here denote derivatives with respect to ζ. On integrating (1.40) and rearranging we readily obtain

$$T(x, t) = \alpha\gamma \int_{x/(2\gamma t)^{1/2}}^{1} e^{\gamma(1-\xi^2)/2} \, d\xi \tag{1.42}$$

and $X(t) = (2\gamma t)^{1/2}$, where the constant γ is determined as the positive root of the transcendental equation

$$\alpha\gamma \int_{0}^{1} e^{\gamma(1-\xi^2)/2} d\xi = 1, \tag{1.43}$$

which we discuss more fully subsequently in this section.

Alternatively, we may derive (1.42) by fixing the boundary with Landau's transformation (see Landau 1950) and employing X as a time variable. With the transformations

$$\rho = \frac{x}{X(t)}, \qquad \tau = X(t), \qquad T(x, t) = u(\rho, \tau), \tag{1.44}$$

we find that (1.22)–(1.24) for β zero becomes

$$\alpha u_{\rho\rho} = u_\rho(1,\ \tau)(\rho u_\rho - \tau u_\tau), \tag{1.45}$$

$$u(0,\ \tau) = 1, \qquad u(1,\ \tau) = 0, \tag{1.46}$$

where subscripts denote partial derivatives, and the arguments of u and its partial derivatives are understood to be $(\rho,\ \tau)$ unless otherwise indicated. Moreover, in the derivation of (1.45) we have utilized (1.24) which becomes

$$u_\rho(1,\ \tau) = -\alpha\tau\frac{d\tau}{dt}, \qquad \tau(0) = 0. \tag{1.47}$$

From these equations it is immediately apparent that a solution for u exists which is a function of ρ only. Thus, with

$$u = A_0(\rho), \qquad \gamma = -A_0'(1)/\alpha, \tag{1.48}$$

where primes here denote differentiation with respect to ρ, we may readily deduce

$$A_0(\rho) = \alpha\gamma \int_\rho^1 e^{\gamma(1-\xi^2)/2}\,d\xi, \tag{1.49}$$

while from (1.47) we again have $\tau = (2\gamma t)^{1/2}$, where γ is determined from (1.43). Thus, we again obtain the exact Neumann solution (1.42).

The advantage of this second approach is that the case $X(0)$ non-zero is also readily deduced. In many practical situations $X(t)$ is generally non-zero initially, especially in a chemical-reaction context where, usually, some reaction has occurred prior to measurements taking place. In this situation, however, the second derivation given still holds except that $(1.47)_2$ is replaced by $X(0) = a$, where a is a constant, and on integration of $(1.47)_1$ we obtain

$$X(t) = (a^2 + 2\gamma t)^{1/2}. \tag{1.50}$$

We also note that in terms of the error function defined by (1.12) Neumann's solution becomes

$$T(x,\ t) = 1 - \frac{\mathrm{erf}(x/2t^{1/2})}{\mathrm{erf}((\gamma/2)^{1/2})}, \qquad X(t) = (2\gamma t)^{1/2}, \tag{1.51}$$

and γ is the positive real root of

$$\alpha\,e^{\gamma/2}(\pi\gamma/2)^{1/2}\,\mathrm{erf}((\gamma/2)^{1/2}) = 1. \tag{1.52}$$

Table 1.2 gives numerical values of roots γ of (1.43) or (1.52) for various values of α. For small α (large γ) we obtain from (1.52), using the standard error function asymptotic expansion (see, for example,

Table 1.2 Numerical values of roots γ of (1.43) for various values of α

α	γ	α	γ	α	γ	α	γ
0.01	6.8520	0.1	3.1600	1.0	0.7690	10.0	0.0968
0.02	5.6718	0.2	2.2459	2.0	0.4321	20.0	0.0492
0.03	5.0028	0.3	1.7810	3.0	0.3012	30.0	0.0330
0.04	4.5405	0.4	1.4875	4.0	0.2313	40.0	0.0248
0.05	4.1902	0.5	1.2819	5.0	0.1878	50.0	0.0199
0.06	3.9101	0.6	1.1286	6.0	0.1581	60.0	0.0166
0.07	3.6780	0.7	1.0093	7.0	0.1365	70.0	0.0142
0.08	3.4808	0.8	0.9136	8.0	0.1201	80.0	0.0125
0.09	3.3100	0.9	0.8349	9.0	0.1072	90.0	0.0111

Carslaw and Jaeger 1965, p. 422),

$$\alpha\left\{e^{\gamma/2}\left(\frac{\pi\gamma}{2}\right)^{1/2} - \sum_{n=0}^{\infty} \frac{(-1)^n (2n)!}{n!\,(2\gamma)^n}\right\} = 1, \tag{1.53}$$

that is

$$e^{\gamma/2}\left(\frac{\pi\gamma}{2}\right)^{1/2} = 1 + \frac{1}{\alpha} + O\left(\frac{1}{\gamma}\right). \tag{1.54}$$

On taking logarithms we obtain the following approximate formula for small α, namely

$$\gamma \simeq 2\log\left(\frac{2}{\pi}\right)^{1/2}\left(1 + \frac{1}{\alpha}\right) - \log\left\{2\log\left(\frac{2}{\pi}\right)^{1/2}\left(1 + \frac{1}{\alpha}\right)\right\}. \tag{1.55}$$

For α large (small γ) we have, from the error function series expansion (again see Carslaw and Jaeger 1965, p. 422),

$$\frac{1}{\alpha} = \gamma \sum_{n=0}^{\infty} \frac{n!\,(2\gamma)^n}{(2n+1)!}, \tag{1.56}$$

that is

$$\frac{1}{\alpha} = \gamma\left(1 + \frac{\gamma}{3} + \frac{\gamma^2}{15} + \frac{\gamma^3}{105} + \dots\right). \tag{1.57}$$

From this equation we obtain

$$\frac{1}{\gamma} = \alpha + \frac{1}{3} - \frac{2}{45\alpha} + \frac{16}{945\alpha^2} + \dots, \tag{1.58}$$

which may be utilized as an approximate formula for large α.

Moreover, for the roots γ of (1.43) we may establish the inequalities

$$\alpha \le \gamma^{-1} \le \alpha + \tfrac{1}{3}, \tag{1.59}$$

which relate to subsequent work on bounding the motion of moving boundaries. In order to establish (1.59) we first observe that γ is positive for α positive. This follows since the integral in (1.43) is positive for any γ. Next, to show $\alpha\gamma \le 1$ we suppose the contrary, namely $\alpha\gamma > 1$, then from (1.57) we have

$$\frac{1}{\alpha\gamma} = \left(1 + \frac{\gamma}{3} + \frac{\gamma^2}{15} + \frac{\gamma^3}{105} + \dots\right) < 1,$$

which is clearly absurd for γ positive, therefore we have established a contradiction and the first inequality of (1.59) follows. For the second inequality of (1.59) we again suppose the converse holds, that is $\gamma^{-1} > \alpha + \tfrac{1}{3}$. Since α is positive it follows that we can, without loss of generality, assume $\gamma < 3$. Now, from the assumption $\gamma^{-1} > \alpha + \tfrac{1}{3}$ we have

$$\frac{\gamma}{(1 - \gamma/3)} < \frac{1}{\alpha},$$

which, from the geometric series (since $0 < \gamma/3 < 1$) and (1.56), gives

$$\sum_{n=0}^{\infty} \left(\frac{\gamma}{3}\right)^n < \sum_{n=0}^{\infty} \frac{n!\,(2\gamma)^n}{(2n + 1)!},$$

which is again clearly absurd for $\gamma > 0$ since evidently

$$\left(\frac{1}{3}\right)^n \ge \frac{n!\,2^n}{(2n + 1)!},$$

with equality for n zero and one, and strict inequality for $n \ge 2$. Thus, a contradiction is again established and the second inequality of (1.59) follows. We note that the $\tfrac{1}{3}$ in (1.59) is the best constant that can be achieved, since from (1.58) it is apparent that γ^{-1} tends to $\alpha + \tfrac{1}{3}$ as α tends to infinity. This and the validity of the inequalities (1.59) are confirmed numerically in Table 1.3. (This proof of (1.59) is due to J. N. Dewynne (1983 private communication).)

To relate (1.51) to subsequent work we note the limiting large α solution. By expansion of (1.42) or (1.49) in powers of γ, and using (1.57), we may show that

$$T(x, t) = (1 - \rho) - \frac{\rho(1 - \rho^2)}{6\alpha} + O\left(\frac{1}{\alpha^2}\right), \tag{1.60}$$

Table 1.3 Confirmation of inequalities (1.59) for various values of α (from Dewynne and Hill (1984))

α	γ^{-1}	$\alpha + \frac{1}{3}$
0.20	0.4453	0.5333
0.25	0.5043	0.5833
0.50	0.7801	0.8333
1.00	1.3005	1.3333
2.00	2.3145	2.3333
5.00	5.3251	5.3333
10.00	10.3291	10.3333
50.00	50.3325	50.3333
100.00	100.3329	100.3333
500.00	500.3332	500.3333

where $\rho = x/X(t)$ and the large α expansion of $X(t)$ becomes

$$X(t) = \left(\frac{2t}{\alpha}\right)^{1/2} - \frac{(t/2)^{1/2}}{3\alpha^{3/2}} + O\left(\frac{1}{\alpha^{5/2}}\right). \tag{1.61}$$

The leading terms correspond to the pseudo-steady-state approximation which is discussed in the following section. The above results are also a special case of the large α expansions which are developed in Chapter 2.

Finally in this section we briefly note some of the numerous extensions of the similarity solution. Kehoe (1972), Knight and Philip (1974), Babu and van Genuchten (1979) and Grundy (1979) all describe similarity solutions for problems involving non-linear diffusivity. Pseudo-similarity solutions of the heat equation for specified moving boundaries are given by Gibson (1958), Langford (1967a), Bluman (1974) and Tait (1979). Exact solutions for both temperature and moisture for a porous media and involving a moving evaporation front are developed by Mikhailov (1975, 1976), Cho (1975) and Lin (1981, 1982). Wilson (1982) utilizes similarity solutions and moving coordinates to analyse a one-dimensional multi-phase Stefan problem with distinct constant densities for each phase. Cross *et al.* (1979) give an analytic solution for the pressure for a simple model of the drying of a porous media. These illustrations by no means exhaust the available literature on generalizations and extensions of the similarity solution.

1.4 Pseudo-steady-state approximations

For heat-diffusion moving-boundary problems by far the most important simple analytic estimate is the pseudo-steady-state approximation. This approximation involves the assumption that the rate of movement of the boundary is very much slower than the rate of diffusion. In other words, the time taken for heat to diffuse a given distance is much smaller than the time taken for complete freezing of the same thickness. In this section we obtain the pseudo-steady-state estimate simply by neglecting the time partial derivative in the diffusion equation. However, in Chapter 2 we show that this approximation emerges as the leading term in a large Stefan number expansion. Thus the pseudo-steady-state estimate applies whenever $\alpha = L/C_1(T_f - T_0)$ is large. That is, either if the specific heat of the solid is small in comparison with the latent heat of freezing or if the coolant temperature T_0 is just below the freezing temperature T_f, so that the difference $T_f - T_0$ is small. In Chapter 3 we show that, for the problems considered here, the pseudo-steady-state temperature over-estimates the actual temperature and the pseudo-steady-state boundary moves in advance of the actual boundary. We first derive the pseudo-steady-state approximations for the planar problem (1.22)–(1.24), the spherical problem (1.30)–(1.32) and the corresponding cylindrical problem. In order to illustrate the procedure further we give two additional examples involving spherical moving boundaries. Both may be viewed as generalizations of (1.30)–(1.32). The first problem in the diffusion context involves not only solid diffusion and external mass transfer at the surface of the spherical particle but, in addition, a first-order chemical reaction on the moving front. In this case the boundary condition $(1.31)_2$ is replaced by

$$T(R(t), t) = \delta \frac{\partial T}{\partial r}(R(t), t), \tag{1.62}$$

where δ is a non-dimensional constant involving the reaction-rate constant (see Bischoff 1965). The second problem involves solid diffusion associated with a rapid chemical reaction and, as well, removal by a second slow chemical reaction, such that the governing Eqn. (1.30) becomes

$$\frac{\partial T}{\partial t} = \frac{\partial^2 T}{\partial r^2} + \frac{2}{r}\frac{\partial T}{\partial r} - k^2 T, \qquad R(t) < r < 1, \tag{1.63}$$

where k is related to the pseudo-first-order rate constant for the slow reaction (see Krishnamurthy and Shah 1979). Finally, we close this section with a brief review of other pseudo-steady-state solutions in the literature.

The pseudo-steady-state estimate for the problem (1.22)–(1.24) is the solution of

$$\frac{\partial^2 T}{\partial x^2} = 0, \qquad 0 < x < X(t), \tag{1.64}$$

satisfying both boundary conditions (1.23). The pseudo-steady-state motion of the boundary is then determined from this solution and (1.24). From (1.64) we have

$$T(x, t) = A(t) + B(t)x, \tag{1.65}$$

where $A(t)$ and $B(t)$ denote functions of time only. On determining these functions such that (1.23) are satisfied, we readily obtain

$$T_{\text{pss}}(x, t) = \frac{X(t) - x}{X(t) + \beta}, \tag{1.66}$$

as the pseudo-steady-state approximation to the temperature profile. From (1.24), (1.66), performing the integration and using the initial condition $X(0) = 0$ we have

$$t_{\text{pss}}(X) = \frac{\alpha}{2} X(X + 2\beta), \tag{1.67}$$

which is the pseudo-steady-state estimate of the motion of the boundary. We observe that for β zero, (1.66) and (1.67) give precisely the leading terms in the large α expansions (1.60) and (1.61), respectively. Figure 1.1 shows, for β zero, the variation of the pseudo-steady-state temperature and the exact temperature given by (1.42). We observe that the pseudo-steady-state estimate is indeed an upper bound to the actual temperature.

For the problem of freezing a liquid in a spherical container, (1.30)–(1.32), the pseudo-steady-state approximation is obtained by solving

$$\frac{\partial^2 T}{\partial r^2} + \frac{2}{r} \frac{\partial T}{\partial r} = 0, \qquad R(t) < r < 1, \tag{1.68}$$

subject to (1.31). From

$$T(r, t) = A(t) + \frac{B(t)}{r}, \tag{1.69}$$

we readily obtain

$$T_{\text{pss}}(r, t) = \frac{1 - [R(t)/r]}{1 + (\beta - 1)R(t)}, \tag{1.70}$$

$$t_{\text{pss}}(R) = \frac{\alpha}{6}(1 - R)[(1 + 2\beta)(1 + R) + 2(\beta - 1)R^2], \tag{1.71}$$

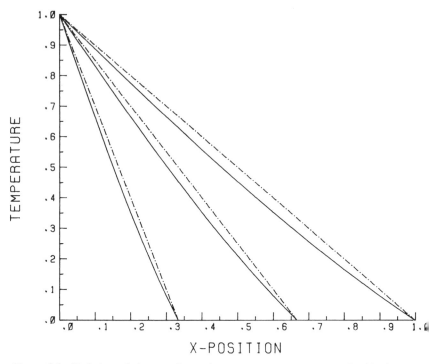

Figure 1.1 Variation of the pseudo-steady-state temperature compared with the exact temperature for a slab with $\alpha = 1$

as the appropriate approximations to the temperature and motion of the boundary. In particular, we observe from (1.71) that the pseudo-steady-state estimate of the time t_c to complete solidification is simply $\alpha(1 + 2\beta)/6$. Similarly, for the problem of freezing a liquid in an infinite circular cylindrical container we deduce from

$$T(r, t) = A(t) + B(t) \log r, \tag{1.72}$$

(1.31) and (1.32) that

$$T_{pss}(r, t) = \frac{\log r - \log R(t)}{\beta - \log R(t)}, \tag{1.73}$$

$$t_{pss}(R) = \frac{\alpha}{4}[2R^2 \log R + (1 + 2\beta)(1 - R^2)], \tag{1.74}$$

are the required approximations. In this case $\alpha(1 + 2\beta)/4$ approximates the time t_c to complete solidification.

The determination of these approximations to the temperature and motion of the boundary is clearly a simple procedure. In the case of a spherical particle with a first-order chemical reaction on the moving front and $(1.31)_2$ replaced by (1.62), we may deduce, in a completely analogous manner,

$$T_{pss}(r, t) = \frac{\delta + R(t) - [R(t)^2/r]}{\delta + R(t) + (\beta - 1)R(t)^2},\qquad(1.75)$$

$$t_{pss}(R) = \frac{\alpha}{6}(1 - R)[6\delta + (1 + 2\beta)(1 + R) + 2(\beta - 1)R^2],\qquad(1.76)$$

so that now $\alpha(1 + 2\beta + 6\delta)/6$ approximates the time to complete solidification. For a spherical particle with solid diffusion and a second slow reaction, so that (1.30) is replaced by (1.63), we obtain

$$T_{pss}(r, t) = \frac{\sinh k[r - R(t)]}{r\{(1 - \beta)\sinh k[1 - R(t)] + \beta k \cosh k[1 - R(t)]\}},\qquad(1.77)$$

$$t_{pss}(R) = \alpha k^{-3}\big\{kR[(1 - \beta)\cosh k(1 - R) + \beta k \sinh k(1 - R)]$$
$$+ [(1 - \beta)\sinh k(1 - R) + \beta k \cosh k(1 - R)] - k\big\},\qquad(1.78)$$

as the appropriate pseudo-steady-state approximations. From (1.78) the pseudo-steady-state estimate of t_c is given by

$$t_{pssc} = \alpha k^{-3}[(1 - \beta)\sinh k + \beta k \cosh k - k].\qquad(1.79)$$

For these two generalizations of (1.30)–(1.32) the reader may confirm that (1.70) and (1.71) both emerge from (1.75), (1.76) and (1.77), (1.78) in the limit of δ and k tending to zero, respectively.

Although the determination of the pseudo-steady-state approximation for the above problems is clearly straightforward, we remark that the process becomes far more complicated for higher-dimensional problems. For example, conformal mapping techniques are employed by Siegel (1968, 1978) and Sproston (1981) for various two-dimensional problems. Siegel (1968) determines the shape of two-dimensional solidified layers formed on one side of a cold surface immersed in a flowing warm liquid. The frozen region is represented by a rectangle in a potential plane and the rectangle is mapped into the physical plane to determine the frozen boundary configuration. The method is illustrated for the problem of solidification on a cold plate of finite width that is insulated along its edges. Siegel (1978) analyses the two-dimensional interface shape of a slab ingot being cast continuously by withdrawing it from a mould. The sides of the ingot are being cooled while the upper boundary of the ingot is in contact with a pool of molten metal. The analysis includes the effect of interface curvature which, for a constant rate of withdrawing the case ingot from the mould, causes the solidification to be non-uniform along the interface. Sproston (1981) also determines the shape of the interface between a solidified layer formed on the inside of a cooled pipe of rectangular cross-section and a warmer flowing liquid. In the steady state convective heat transfer from the liquid to the interface balances the heat transfer by conduction through the solidified layer. For pipes of various aspect ratio, the heat transfer is computed as a function of a non-dimensional parameter involving the interfacial convection coefficient and the thermal conductivity of the layer, and it is found that the extent of solidification displays a critical thickness characteristic. Both Duda *et al.* (1975) and Kern and Hansen (1976) utilize a pseudo-steady-state approximation for two-dimensional problems involving a cylindrical geometry. The first paper also presents a numerical technique for unsteady two-dimensional diffusion problems involving moving boundaries. The second paper relates to an interesting application involved in the temporary storage of material in cylindrical containers or silos. In general, the analysis of the pseudo-steady-state approximation for two- and three-dimensional problems is difficult and non-trivial.

1.5 Review of other analytic methods

In this section we briefly review the most important analytic approximations not discussed in subsequent chapters. There are other special approximate methods which all have numerous extensions and applications in the literature. However, we mention here only those associated with the names of Biot, Boley, Selim, Seagrave and Tao. Biot's variational method is based on the concepts of irreversible thermodynamics and has been applied to a number of problems. The general theory is described by Biot (1957, 1959). Concepts of thermal potential, dissipation function and generalized thermal force are introduced which lead to ordinary differential equations of the Lagrangian type for the thermal flow field. The approach is completely analogous to dynamics although, because of the particular nature of heat-flow phenomena, special procedures are developed to formulate each problem in the simplest way. The method although applicable to a wide variety of heat-flow problems, including inhomogeneous and non-linear problems, depends on an assumed form for the temperature. Biot and Daughaday (1962) employ a cubic temperature profile for the problem of a half-space subjected to a constant rate of heat input at a melting surface. Lardner (1967) utilizes a quadratic temperature profile for the classical Stefan problem and a comparison of the approximation with the exact Neumann solution is made. Zyszkowski (1969) employs Biot's method to determine the transient temperature for one-dimensional problems with internal heat generation and non-linear boundary conditions for both Newton cooling and black-body radiation. Prasad and Agrawal (1972, 1974) use simple linear temperature profiles for classical melting problems. Chung and Yeh (1975) and Yeh and Chung (1977, 1979) describe Biot's technique for phase-change problems of materials also subject to convection and radiation and with variable thermal properties. Further references to other applications of Biot's variational principle can be found in the above papers. The following three methods differ from Biot's approach in that they may be considered 'exact' in the sense that all three involve various series representations of the actual solution. In principle, any number of terms of the series may be calculated. However, in practice, only the first few terms are usually obtained and the resulting terminated series is an approximation to the solution.

Boley (1961) introduces the 'embedding technique' for one-dimensional problems and obtains (Boley 1968) a general one-dimensional starting solution for melting and solidifying slabs. In this approach the melting or solidifying body of changing dimensions is mathematically extended to occupy the space filled by the actual body before change of phase. An unknown fictitious heat input is applied at the boundary of the extended body, whose magnitude is then determined in such a manner as to satisfy the actual boundary conditions on the moving liquid–solid interface. In this way the original partial differential boundary-value problem is replaced by an integro-differential boundary-value problem which may be solved either in series form or numerically. The principal advantage of this method is that it allows an explicit expression for the temperature in terms of the fictitious heat input. Boley and Yagoda (1969) extend the technique to two-dimensional regions and obtain a short-time solution for melting half-space problems. With a general heat input the problems considered are such that melting starts at a point on the surface and spreads both along the surface and toward the interior of the body. It is found that the shape of the melt interface is independent of both the applied heat input and material properties. Moreover, initial melt propagation along and normal to the surface is proportional to $t^{1/2}$ and $t^{3/2}$, respectively, where time t is measured from the start of melting. Results obtained in this paper are simplified and extended to axially symmetric problems by Yagoda and Boley (1970). Lederman and Boley (1970) utilize the embedding technique for the problem of melting a cylinder under arbitrary heat inputs applied on the outer radius. Analytical solutions valid for short times and numerical solutions for all times are obtained. Although the analysis involved in this procedure is complicated the end results appear both simple and significant.

Selim and Seagrave (1973a, 1973b) develop finite integral transform techniques for the solution of plane, cylindrical and spherical moving-boundary problems. Use of the finite integral transform removes the space variable from the partial differential equation and the resulting system of ordinary differential equations are integrated numerically. Selim and Seagrave (1973a) use the method to obtain solutions to the problem of freezing or melting a finite slab with boundary conditions either prescribed temperature, prescribed heat flux or Newton cooling. Selim and Seagrave (1973b) extend this method to the classical problems of inward freezing of cylinders and spheres with prescribed constant surface temperature. The technique, although requiring considerable computing, appears to be accurate and effective. In the final paper of the series (Selim and Seagrave 1973c) the results obtained for the spherical moving boundary are utilized to determine diffusivities from experimental data for the diffusion-controlled penetration of a reactant into a spherical ion-exchange particle.

The exact Neumann solutions apply only to the case of constant initial temperature and prescribed constant surface temperature. In a sequence of papers, Tao gives formal series solutions to a number of classical moving-boundary problems with non-constant initial and boundary conditions. The series solutions generally involve the complementary error function and its integrals, and constitute one representation of the exact solution. The existence of such series solutions does not preclude other representations of the exact solution which are possibly simpler in form. Tao (1978) gives a series solution of the classical Stefan problem for a half-space with arbitrary initial and surface temperatures. The solutions involve functions and polynomials of the similarity variable $x/t^{1/2}$ and time t, and the convergence of the series is established. Tao (1979) treats the same problem but with an induced motion due to density changes. The non-linear diffusion equation including this effect of the density change is transformed to a diffusion equation without the density-change term. The moving boundary is shown to be of order 'square root time' for small time according to whether or not the initial temperature at the interface is discontinuous or continuous. Tao (1980) similarly gives series solutions to the problem of solidification of a binary alloy in a semi-infinite region, with arbitrary initial temperature and composition and arbitrary surface temperature. Tao (1981, 1982a) extends the technique to the problems of a semi-infinite material with an arbitrary prescribed heat flux and with an imperfect thermal contact at the moving front, respectively. Tao (1982b) considers the classical problem of solidification of two semi-infinite materials but with arbitrary initial conditions. Rubinstein (1971) refers to this problem as the Cauchy–Stefan problem and Tao (1982b) shows that, depending on the prescribed initial conditions, there are four possibilities:

(i) Solidification starts immediately, with 'square root time' boundary movement for small time.
(ii) Solidification starts immediately but the boundary moves as $t^{n/2}$ for small time, where $n \geq 2$.
(iii) No solidification ever occurs.
(iv) There is a pre-solidification period during which the temperatures of both phases are redistributed and solidification occurs only after the interface acquires the freezing temperature.

1.6 Numerical solution of classical Stefan problems

In this section we briefly summarize the main qualitative features of temperature profiles and boundary motions as obtained from the explicit

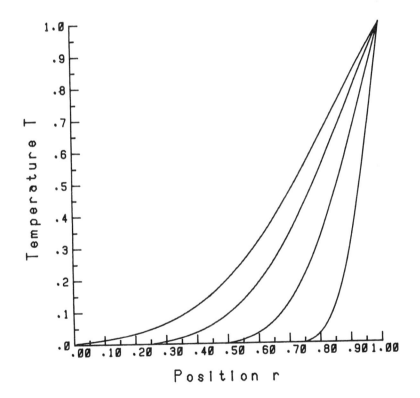

Figure 1.2 Variation in temperature from a numerical enthalpy scheme for spherical solidification with $\alpha = 0.01$ and β zero

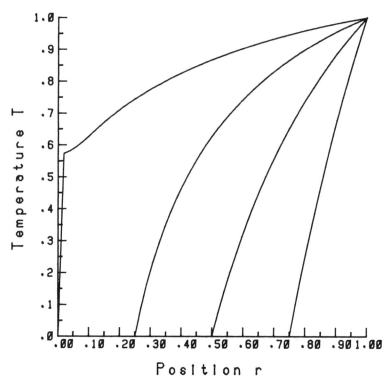

Figure 1.3 Variation in temperature from a numerical enthalpy scheme for spherical solidification with $\alpha = 10.0$ and β zero

numerical enthalpy scheme of Voller and Cross (1981). For the enthalpy method in general we refer the reader to Elliott and Ockendon (1982). The particular implementation of Voller and Cross (1981) converges rapidly for small α but, because of rounding error, is not particularly efficient for large α. The radiation condition is implemented by assuming the partial differential equation holds on the surface and solving in a slightly larger domain, so that the normal derivative may be obtained by a two-sided procedure. Although Figs 1.2–1.5 all describe the problem of spherical solidification, the corresponding curves for cylindrical solidification have similar characteristics.

Figures 1.2 and 1.3 show temperature profiles for small and large α, respectively, and β zero. We observe that for small α the profiles are essentially concave upwards, while for large α the temperature curves are concave downwards. As noted previously, the velocity of the boundary decreases with increasing α and the concavity of the temperature profiles appears to hinge on the magnitude of the boundary velocity, namely

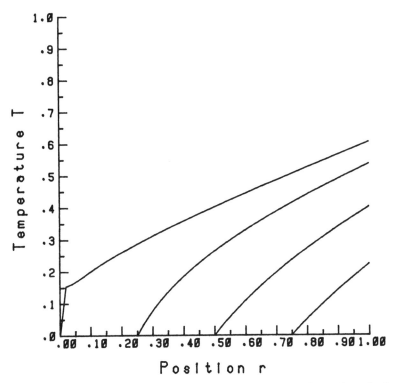

Figure 1.4 Variation in temperature from a numerical enthalpy scheme for spherical solidification with α and β both unity

$V(t) = -dR/dt$. This may be seen by differentiating $(1.31)_2$ with respect to time and using (1.32) and (1.34) to give

$$\frac{\partial^2 T}{\partial r^2}(R(t), t) = \alpha V\left(V - \frac{\lambda}{R}\right), \qquad (1.80)$$

which is the determining equation for the concavity at the moving boundary. Thus, for small α, V is large and the second derivative is positive except at the point of complete solidification. On the other hand, if α is large, V is small and the second term dominates (1.80), and the temperature profiles are concave downwards. The boundary-layer phenomenon for large α, discussed in the following chapter, is clearly apparent from Fig. 1.3. Figure 1.4 shows temperature profiles for an intermediate value of α and β non-zero. For β non-zero the characteristics are similar to β zero except, of course, that the surface temperature is no longer fixed. For both spherical and cylindrical solidification there is little variability in the boundary motions for varying α and β. Figure 1.5

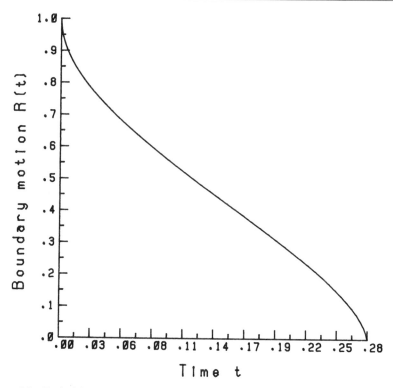

Figure 1.5 Typical boundary motion from a numerical enthalpy scheme for spherical solidification with $\alpha = 1.0$ and β zero

shows a typical boundary motion. These results appear to indicate that the boundary velocity is infinite at complete solidification and that for sufficiently large α and β the boundary motion becomes strictly concave downwards. Finally we note that since T is zero initially we may deduce from the surface and Stefan conditions that for the three basic problems the boundary velocity is initially $(\alpha\beta)^{-1}$, Thus the problems with a prescribed surface temperature and no heat loss $(\beta = 0)$ all have an infinite initial boundary velocity.

1.7 Additional symbols used

$A_0(\rho)$ temperature for Neumann's exact solution, (1.48)
$A(t)$, $B(t)$ arbitrary functions of time, (1.65)
k related to pseudo-first-order rate constant, (1.63)
$\bar{t}$, $\bar{T}$, $\bar{x}$, $\bar{X}$ one-parameter group transformation of time, temperature, position and boundary position, respectively, (1.36)

$u(\rho, \tau)$	temperature for Neumann's exact solution, (1.44)
$V(t)$	magnitude of boundary velocity, (1.80)

Greek symbols

δ	related to reaction-rate constant, (1.62)
ε	arbitrary parameter, (1.36)
ζ	similarity variable $x/t^{1/2}$, (1.40)
ρ	slab variable $x/X(t)$, (1.44)
τ	moving-boundary variable $X(t)$, (1.44)
ϕ	temperature as function of $x/t^{1/2}$, (1.39)
Φ	functional form of temperature, (1.37)
Ψ	functional form of moving boundary, (1.37)

2

Large Stefan number expansion

2.1 Introduction

In the remainder of the book we focus primarily on the three problems of freezing an infinite slab, freezing a liquid in a spherical container and freezing a liquid in an infinite circular cylindrical container, all of which are assumed initially at their uniform fusion temperature. For convenience we give here the non-dimensional formulations of these three problems. For the semi-infinite half-space $(0, \infty)$ with Newtonian cooling on the surface $x = 0$, the problem becomes the determination of the non-dimensional temperature $T(x, t)$ and moving boundary $X(t)$ such that

$$\frac{\partial T}{\partial t} = \frac{\partial^2 T}{\partial x^2}, \qquad 0 < x < X(t), \tag{2.1}$$

$$T(0, t) - \beta \frac{\partial T}{\partial x}(0, t) = 1, \qquad T(X(t), t) = 0, \tag{2.2}$$

$$\frac{\partial T}{\partial x}(X(t), t) = -\alpha \frac{dX}{dt}, \qquad X(0) = 0, \tag{2.3}$$

where the constants α and β and variables are defined in terms of the physical parameters of the problem by (1.18)–(1.20). For the spherical container the problem becomes the determination of $T(r, t)$ and $R(t)$ such that

$$\frac{\partial T}{\partial t} = \frac{\partial^2 T}{\partial r^2} + \frac{2}{r}\frac{\partial T}{\partial r}, \qquad R(t) < r < 1, \tag{2.4}$$

$$T(1, t) + \beta \frac{\partial T}{\partial r}(1, t) = 1, \qquad T(R(t), t) = 0, \tag{2.5}$$

$$\frac{\partial T}{\partial r}(R(t), t) = -\alpha \frac{dR}{dt}, \qquad R(0) = 1. \tag{2.6}$$

For the infinite cylindrical container, Eqs (2.5) and (2.6) still hold, while (2.4) is replaced by

$$\frac{\partial T}{\partial t} = \frac{\partial^2 T}{\partial r^2} + \frac{1}{r}\frac{\partial T}{\partial r}, \qquad R(t) < r < 1, \tag{2.7}$$

and for both the sphere and cylinder the non-dimensional variables are defined by (1.28) and (1.29). We remark that for all three problems the temperature is normalized such that $0 \le T \le 1$, while β is a non-negative constant such that β zero corresponds to the case of no Newton heat loss on the surface and the surface is maintained at a uniform low temperature. The Stefan number $\alpha = L/C_1(T_f - T_0)$ is strictly positive and represents the ratio of freezing times to diffusion times. In this chapter we give the solutions of the above problems applicable for large α. That is, these solutions apply whenever the times taken for the solidified phase to acquire a given thickness are very much greater than the times for heat to diffuse the same distance. Thus, for example, large Stefan number solutions are generally applicable for ice-formation problems.

For completeness we present, in the following section, the first derivations of large α solutions. This method relates to general solutions of the heat-diffusion equation and was first given for the plane by Stefan (1889c), for the sphere by Bischoff (1963) and for the cylinder by Pekeris and Slichter (1939). These authors employed only the first few terms of general solutions noted by Stefan (1889c) for the plane and Langford (1967b) for the sphere and cylinder. In Sections 2.3–2.5 we present more formal derivations of large α solutions based on the following boundary fixing transformations:

$$\rho = \frac{x}{X(t)} \quad \text{(plane)}, \tag{2.8}$$

$$\rho = \frac{1-r}{1-R(t)} \quad \text{(sphere)}, \tag{2.9}$$

$$\rho = \frac{\log r}{\log R(t)} \quad \text{(cylinder)}, \tag{2.10}$$

for which all three map the variable region occupied by the solidified phase into the fixed unit interval $0 \le \rho \le 1$. The first two transformations are essentially Landau's transformation (see Landau 1950), while the third appears to have been first employed by Huang and Shih (1975b). The approach adopted in these sections is essentially due to Huang and Shih (1975a, 1975b), although several other authors have made similar contributions. These developments and the general perturbation literature relating to these problems are briefly reviewed in the final section of the chapter. Finally in this section we detail the general solutions due to Stefan (1889c) and Langford (1967b) which are utilized in the following section.

These solutions apply when both the non-dimensional temperature and, essentially, the heat-flow rate on some fixed boundary are pre-scribed functions of time, $A(t)$ and $B(t)$, respectively. Thus, with x_0 and r_0 denoting an arbitrary fixed plane and spherical or cylindrical surface, we have

$$T(x_0, t) = A(t), \qquad \frac{\partial T}{\partial x}(x_0, t) = B(t) \quad \text{(plane)}, \tag{2.11}$$

$$T(r_0, t) = A(t), \qquad \lim_{r \to r_0}\left(r^2 \frac{\partial T}{\partial r}(r, t)\right) = -B(t) \quad \text{(sphere)}, \tag{2.12}$$

$$T(r_0, t) = A(t), \qquad \lim_{r \to r_0}\left(r \frac{\partial T}{\partial r}(r, t)\right) = B(t) \quad \text{(cylinder)}. \tag{2.13}$$

It is convenient to give the solutions corresponding to x_0 (r_0) zero and non-zero separately. With the notation

$$A^{(n)}(t) = \frac{d^n A(t)}{dt^n}, \qquad B^{(n)}(t) = \frac{d^n B(t)}{dt^n}, \tag{2.14}$$

we have for x_0 (r_0) zero

$$T(x, t) = \sum_{n=0}^{\infty} \left\{ \frac{A^{(n)}(t)x^{2n}}{(2n)!} + \frac{B^{(n)}(t)x^{2n+1}}{(2n+1)!} \right\} \quad \text{(plane)}, \tag{2.15}$$

$$T(r, t) = \sum_{n=0}^{\infty} \left\{ \frac{A^{(n)}(t)r^{2n}}{(2n+1)!} + \frac{B^{(n)}(t)r^{2n-1}}{(2n)!} \right\} \quad \text{(sphere)}, \tag{2.16}$$

$$T(r, t) = \sum_{n=0}^{\infty} \left\{ A^{(n)}(t) + B^{(n)}(t)\left[\log r - \sum_{j=1}^{n} \frac{1}{j}\right] \right\} \frac{(r/2)^{2n}}{n!\,n!} \quad \text{(cylinder)}, \tag{2.17}$$

where the term in the square brackets for $n = 0$ is taken to be simply $\log r$. By direct differentiation it is a simple matter to verify that the above are indeed solutions of the heat-diffusion equation. Langford (1967b) notes that these solutions are general in the sense that special choices of $A(t)$ and $B(t)$ yield other solutions of the heat equation. For example, the classical Fourier and Bessel series solutions may be obtained in this way. The $n = 0, 1$ terms of the above series, which are used in the next section to determine large α solutions, are

$$T(x, t) = A(t) + B(t)x + \left\{ \frac{dA}{dt}\frac{x^2}{2} + \frac{dB}{dt}\frac{x^3}{6} \right\} + \ldots \quad \text{(plane)}, \tag{2.18}$$

$$T(r, t) = A(t) + \frac{B(t)}{r} + \left\{ \frac{dA}{dt} \frac{r^2}{6} + \frac{dB}{dt} \frac{r}{2} \right\} + \ldots \quad \text{(sphere)}, \qquad (2.19)$$

$$T(r, t) = A(t) + B(t) \log r + \left\{ \frac{dA}{dt} - \frac{dB}{dt} + \frac{dB}{dt} \log r \right\} \frac{r^2}{4} + \ldots \quad \text{(cylinder)},$$
$$(2.20)$$

and we note that the leading terms take the same form as the pseudo-steady-state solutions (1.65), (1.69) and (1.72), respectively.

If x_0 and r_0 are non-zero then solutions satisfying (2.11)–(2.13) are given by

$$T(x, t) = \sum_{n=0}^{\infty} \left\{ A^{(n)}(t) \frac{(x - x_0)^{2n}}{(2n)!} + B^{(n)}(t) \frac{(x - x_0)^{2n+1}}{(2n + 1)!} \right\} \quad \text{(plane)},$$
$$(2.21)$$

$$T(r, t) = \sum_{n=0}^{\infty} \left\{ \frac{A^{(n)}(t)}{r} \frac{(r - r_0)^{2n}}{(2n + 1)!} (r + 2nr_0) - \frac{B^{(n)}(t)}{rr_0} \frac{(r - r_0)^{2n+1}}{(2n + 1)!} \right\}$$
$$\text{(sphere)}, \quad (2.22)$$

$$T(r, t) = \sum_{n=0}^{\infty} \left\{ A^{(n)}(t) c_n\left(\frac{r^2}{4}, \frac{r_0^2}{4}\right) + \frac{B^{(n)}(t)}{2} e_n\left(\frac{r^2}{4}, \frac{r_0^2}{4}\right) \right\} \quad \text{(cylinder)},$$
$$(2.23)$$

where the functions $c_n(z, z_0)$ and $e_n(z, z_0)$ do not admit simple expressions. Details for these functions for arbitrary n can be found in Appendix 1. With $z = r^2/4$ and $z_0 = r_0^2/4$ the first four of each are given explicitly by

$$c_0(z, z_0) = 1, \qquad c_1(z, z_0) = (z - z_0) - z_0 \log\left(\frac{z}{z_0}\right),$$

$$c_2(z, z_0) = \frac{z - z_0}{4} (z + 5z_0) - \frac{z_0}{2} (z_0 + 2z) \log\left(\frac{z}{z_0}\right), \qquad (2.24)$$

$$c_3(z, z_0) = \frac{z - z_0}{36} (z^2 + 19zz_0 + 10z_0^2) - \frac{z_0}{12} (z_0^2 + 6z_0 z + 3z^2) \log\left(\frac{z}{z_0}\right),$$

and

$$e_0(z, z_0) = \log\left(\frac{z}{z_0}\right), \qquad e_1(z, z_0) = 2(z_0 - z) + (z + z_0) \log\left(\frac{z}{z_0}\right),$$

$$e_2(z, z_0) = \tfrac{3}{4}(z_0 - z)(z_0 + z) + \tfrac{1}{4}(z^2 + 4zz_0 + z_0^2) \log\left(\frac{z}{z_0}\right), \qquad (2.25)$$

$$e_3(z, z_0) = \frac{z_0 - z}{108} (11z_0^2 + 38z_0 z + 11z^2) + \frac{z + z_0}{36} (z^2 + 8zz_0 + z_0^2) \log\left(\frac{z}{z_0}\right),$$

and the reader is referred to Appendix 1 and Langford (1967b) for further details. We note that these functions also arise in the formal series solution for cylindrical freezing given in Chapter 4. We also remark that the solutions (2.21), (2.22) and (2.17) are utilized to deduce simple closed-form exact solutions of phase-change problems with boundaries moving at constant velocity (see Langford 1966, 1967a).

2.2 Derivation from general solutions

The large α solutions obtained by Stefan (1889c), Bischoff (1963) and Pekeris and Slichter (1939) for plane, spherical and cylindrical geometries, respectively, and for no Newton cooling (that is, β zero) are essentially obtained by stretching the time variable, thus

$$\bar{t} = \frac{t}{\alpha},$$

$$(2.26)$$

and assuming the following expansions for the functions $A(t)$ and $B(t)$ in (2.18)–(2.20):

$$A(t) = a_0(\bar{t}) + \frac{a_1(\bar{t})}{\alpha} + O\left(\frac{1}{\alpha^2}\right),$$
$$B(t) = b_0(\bar{t}) + \frac{b_1(\bar{t})}{\alpha} + O\left(\frac{1}{\alpha^2}\right).$$

$$(2.27)$$

Moreover, the plane boundary $X(t)$ and spherical and cylindrical boundaries $R(t)$ are assumed to admit expansions of the form

$$X(t) = X_0(\bar{t}) + \frac{X_1(\bar{t})}{\alpha} + O\left(\frac{1}{\alpha^2}\right),$$

$$(2.28)$$

$$R(t) = R_0(\bar{t}) + \frac{R_1(\bar{t})}{\alpha} + O\left(\frac{1}{\alpha^2}\right).$$

$$(2.29)$$

The large Stefan number solutions now follow from the first two terms of the general solutions (that is, (2.18)–(2.20)), and (2.2), (2.3) for plane geometry and (2.5), (2.6) for either spherical or cylindrical geometry. Details for the three geometries now follow.

2.2.1 Plane freezing. From the above expansions, (2.2), (2.3) and (2.18), and noting that with the stretched time variable $\bar{t}$ defined by (2.26) the velocity term in (2.3) is order one while the term involving time derivatives in (2.18) is order α^{-1}, we obtain the following system of equations:

$$\left(a_0 + \frac{a_1}{\alpha}\right) - \beta\left(b_0 + \frac{b_1}{\alpha}\right) = 1,$$

$$\left(a_0 + \frac{a_1}{\alpha}\right) + \left(b_0 + \frac{b_1}{\alpha}\right)\left(X_0 + \frac{X_1}{\alpha}\right) + \frac{1}{\alpha}\left(\dot{a}_0 \frac{X_0^2}{2} + \dot{b}_0 \frac{X_0^3}{6}\right) = 0, \qquad (2.30)$$

$$\left(b_0 + \frac{b_1}{\alpha}\right) + \frac{1}{\alpha}\left(\dot{a}_0 X_0 + \dot{b}_0 \frac{X_0^2}{2}\right) = -\left(\dot{X}_0 + \frac{\dot{X}_1}{\alpha}\right),$$

where the dot here denotes differentiation with respect to $\bar{t}$ and higher-order terms in α^{-1} have been neglected. From the order one terms, performing an integration we readily obtain

$$a_0(\bar{t}) = \frac{X_0}{X_0 + \beta}, \qquad b_0(\bar{t}) = \frac{-1}{X_0 + \beta}, \qquad \bar{t} = \tfrac{1}{2}X_0(X_0 + 2\beta), \qquad (2.31)$$

which is simply the pseudo-steady-state approximation given by (1.66) and (1.67). On using these expressions and eliminating derivatives with respect to $\bar{t}$ by means of the relation

$$\dot{X}_0 = (X_0 + \beta)^{-1}, \qquad (2.32)$$

we may deduce from the order α^{-1} terms of (2.30) the following first-order equation for X_1 as a function of X_0, that is

$$(X_0 + \beta)\frac{dX_1}{dX_0} + X_1 = -\frac{X_0^3}{3(X_0 + \beta)^2} - \frac{\beta X_0}{(X_0 + \beta)}. \qquad (2.33)$$

On integration, and using X_1 zero when X_0 zero, we readily obtain

$$X_1(\bar{t}) = \frac{-X_0^2(X_0 + 3\beta)}{6(X_0 + \beta)^2}, \qquad (2.34)$$

while $a_1(\bar{t})$ and $b_1(\bar{t})$ are found to be given by

$$a_1(\bar{t}) = -\frac{\beta X_0^2(X_0 + 3\beta)}{3(X_0 + \beta)^4}, \qquad b_1(\bar{t}) = -\frac{X_0^2(X_0 + 3\beta)}{3(X_0 + \beta)^4}. \qquad (2.35)$$

Altogether from (2.26), (2.28), (2.31)$_3$ and (2.34) we obtain for the motion of the boundary

$$X(t) = (\beta^2 + 2t/\alpha)^{1/2} - \beta - \frac{[\beta^3 - (\beta^2 - t/\alpha)(\beta^2 + 2t/\alpha)^{1/2}]}{3\alpha(\beta^2 + 2t/\alpha)} + O\left(\frac{1}{\alpha^2}\right), \qquad (2.36)$$

the first two terms of which coincide with (1.61) for β zero. Altogether for the temperature we have from the above equations

$$T(x, t) = \frac{X_0 - x}{X_0 + \beta} + \frac{x^2(x + 3\beta)(X_0 + \beta) - 2X_0^2(x + \beta)(X_0 + 3\beta)}{6\alpha(X_0 + \beta)^4} + O\left(\frac{1}{\alpha^2}\right),$$

(2.37)

which again may be reconciled with (1.60) for the case β zero, on noting that in this case

$$\rho = \left(1 - \frac{1}{6\alpha}\right)\frac{x}{X_0} + O\left(\frac{1}{\alpha^2}\right).$$

(2.38)

Finally, for plane freezing we note that from the relation

$$X = X_0 - \frac{X_0^2(X_0 + 3\beta)}{6\alpha(X_0 + \beta)^2} + O\left(\frac{1}{\alpha^2}\right),$$

(2.39)

we may deduce

$$X_0 = X + \frac{X^2(X + 3\beta)}{6\alpha(X + \beta)^2} + O\left(\frac{1}{\alpha^2}\right),$$

(2.40)

and therefore from (2.26) and (2.31)$_3$ we have

$$t = \frac{\alpha X}{2}(X + 2\beta) + \frac{X^2(X + 3\beta)}{6(X + \beta)} + O\left(\frac{1}{\alpha}\right),$$

(2.41)

which is the form of the large α motion of the boundary referred to most frequently in subsequent work. We shall loosely refer to the first two terms of (2.41) as the 'order one corrected motion'. We establish later that these two terms constitute an upper bound to the actual boundary motion $t(X)$.

2.2.2 Spherical freezing.

From (2.19), (2.26), (2.27) and (2.29) we may deduce the following system of equations from (2.5) and (2.6):

$$\left(a_0 + \frac{a_1}{\alpha}\right) + (1 - \beta)\left(b_0 + \frac{b_1}{\alpha}\right) + \frac{1}{\alpha}\left(\dot{a}_0\frac{(1 + 2\beta)}{6} + \dot{b}_0\frac{(1 + \beta)}{2}\right) = 1,$$

$$\left(a_0 + \frac{a_1}{\alpha}\right) + \frac{1}{R_0}\left(b_0 + \frac{b_1}{\alpha}\right)\left(1 - \frac{R_1}{\alpha R_0}\right) + \frac{1}{\alpha}\left(\dot{a}_0\frac{R_0^2}{6} + \dot{b}_0\frac{R_0}{2}\right) = 0, \quad (2.42)$$

$$-\frac{1}{R_0^2}\left(b_0 + \frac{b_1}{\alpha}\right)\left(1 - \frac{2R_1}{\alpha R_0}\right) + \frac{1}{\alpha}\left(\dot{a}_0\frac{R_0}{3} + \frac{\dot{b}_0}{2}\right) = -\left(\dot{R}_0 + \frac{\dot{R}_1}{\alpha}\right),$$

where again the dot denotes differentiation with respect to $\bar{t}$. The order one system of equations yields

$$a_0(\bar{t}) = \frac{1}{1 + (\beta - 1)R_0}, \qquad b_0(\bar{t}) = \frac{-R_0}{1 + (\beta - 1)R_0}, \tag{2.43}$$

and

$$\dot{R}_0 = \frac{-1}{R_0[1 + (\beta - 1)R_0]}, \tag{2.44}$$

which on integration, using $R_0(0) = 1$, gives the pseudo-steady-state motion (1.71), namely

$$\bar{t} = \tfrac{1}{6}(1 - R_0)[(1 + 2\beta)(1 + R_0) + 2(\beta - 1)R_0^2]. \tag{2.45}$$

Using (2.43) the order α^{-1} system of equations for $a_1(\bar{t})$, $b_1(\bar{t})$ and $R_1(\bar{t})$ becomes

$$a_1 - (\beta - 1)b_1 = -\frac{1 + \beta + \beta^2}{3R_0[1 + (\beta - 1)R_0]^3},$$

$$a_1 + \frac{b_1}{R_0} = -\frac{R_1}{R_0[1 + (\beta - 1)R_0]} - \frac{3 + (\beta - 1)R_0}{6[1 + (\beta - 1)R_0]^3}, \tag{2.46}$$

$$-\frac{b_1}{R_0^2} - \frac{2R_1}{R_0^2[1 + (\beta - 1)R_0]} + \frac{3 + 2(\beta - 1)R_0}{6R_0[1 + (\beta - 1)R_0]^3} = -\dot{R}_1.$$

On eliminating a_1, b_1 and the derivative with respect to $\bar{t}$ by means of (2.44) we eventually obtain the following first-order differential equation for R_1 as a function of R_0:

$$R_0[1 + (\beta - 1)R_0]\frac{dR_1}{dR_0} + [1 + 2(\beta - 1)R_0]R_1$$

$$= \frac{1}{3(\beta - 1)}\left\{[1 + (\beta - 1)R_0] - \frac{\beta^3}{[1 + (\beta - 1)R_0]^2}\right\}, \tag{2.47}$$

which, using R_1 zero when R_0 is unity, integrates to give

$$R_1(\bar{t}) = \frac{(1 - R_0)^2[1 + 2\beta + (\beta - 1)R_0]}{6R_0[1 + (\beta - 1)R_0]^2}. \tag{2.48}$$

Thus we have

$$R = R_0 + \frac{(1 - R_0)^2[1 + 2\beta + (\beta - 1)R_0]}{6\alpha R_0[1 + (\beta - 1)R_0]^2} + O\left(\frac{1}{\alpha^2}\right), \tag{2.49}$$

and therefore

$$R_0 = R - \frac{(1 - R)^2[1 + 2\beta + (\beta - 1)R]}{6\alpha R[1 + (\beta - 1)R]^2} + O\left(\frac{1}{\alpha^2}\right). \tag{2.50}$$

From this equation, (2.26) and (2.45) we may deduce the following order one corrected motion,

$$t = \frac{\alpha}{6}(1 - R)[(1 + 2\beta)(1 + R) + 2(\beta - 1)R^2]$$
$$+ \frac{(1 - R)^2[1 + 2\beta + (\beta - 1)R]}{6[1 + (\beta - 1)R]} + O\left(\frac{1}{\alpha}\right), \tag{2.51}$$

and we observe that although (2.49) and (2.50) are singular as the boundary approaches the centre, remarkably the order one correction in (2.51) remains finite. The final expression for the temperature up to order α^{-1} becomes

$$T(r, t) = \frac{1 - R_0/r}{1 + (\beta - 1)R_0} + \frac{1}{6\alpha}\left\{2(1 + \beta + \beta^2)\left(\frac{1}{r} - \frac{1}{R_0}\right)\right.$$
$$+ \frac{r}{R_0}[3 + (\beta - 1)r][1 + (\beta - 1)R_0]$$
$$- \frac{R_0}{r}[3 + (\beta - 1)R_0][1 + (\beta - 1)r]$$
$$- \frac{(1 - R_0)^2}{rR_0}[1 + 2\beta + (\beta - 1)R_0][1 + (\beta - 1)r]\right\} \Bigg/$$
$$[1 + (\beta - 1)R_0]^4 + O\left(\frac{1}{\alpha^2}\right), \tag{2.52}$$

which appears not to simplify further. However, we may readily check that the given expression satisfies the boundary conditions (2.5) up to order α^{-1}. We may also confirm that the analysis presented here includes that given by Bischoff (1963) for the case β zero.

2.2.3 Cylindrical freezing.
From the solution (2.20) for the cylinder and the assumed expansions (2.27) and (2.29) for the functions $A(t)$, $B(t)$ and $R(t)$, respectively, we obtain the following system of equations from (2.5) and (2.6):

$$\left(a_0 + \frac{a_1}{\alpha}\right) + \beta\left(b_0 + \frac{b_1}{\alpha}\right) + \frac{1}{4\alpha}\left(\dot{a}_0(1 + 2\beta) - \dot{b}_0(1 + \beta)\right) = 1,$$

$$\left(a_0 + \frac{a_1}{\alpha}\right) + \left(b_0 + \frac{b_1}{\alpha}\right)\left(\log R_0 + \frac{R_1}{\alpha R_0}\right) + \frac{R_0^2}{4\alpha}\left(\dot{a}_0 - \dot{b}_0 + \dot{b}_0 \log R_0\right) = 0,$$

$$\frac{1}{R_0}\left(b_0 + \frac{b_1}{\alpha}\right)\left(1 - \frac{R_1}{\alpha R_0}\right) + \frac{R_0}{4\alpha}\left(2\dot{a}_0 - \dot{b}_0 + 2\dot{b}_0 \log R_0\right) = -\left(\dot{R}_0 + \frac{\dot{R}_1}{\alpha}\right).$$
$$\tag{2.53}$$

On solving the order one system we obtain

$$a_0(\bar{t}) = -\frac{\log R_0}{\beta - \log R_0}, \qquad b_0(\bar{t}) = \frac{1}{\beta - \log R_0}, \qquad (2.54)$$

and

$$\dot{R}_0(\bar{t}) = -\frac{1}{R_0(\beta - \log R_0)}, \qquad (2.55)$$

which on integration using the initial condition $R_0(0)$ is unity gives the pseudo-steady-state motion

$$\bar{t} = \tfrac{1}{4}[2R_0^2 \log R_0 + (1 + 2\beta)(1 - R_0^2)]. \qquad (2.56)$$

From these results the system of equations to determine $a_1(\bar{t})$, $b_1(\bar{t})$ and $R_1(\bar{t})$ becomes

$$a_1 + \beta b_1 = -\frac{1 + 2\beta + 2\beta^2}{4R_0^2(\beta - \log R_0)^3},$$

$$a_1 + b_1 \log R_0 = -\frac{R_1}{R_0(\beta - \log R_0)} - \frac{\beta + 1 - \log R_0}{4(\beta - \log R_0)^3}, \qquad (2.57)$$

$$\frac{b_1}{R_0} - \frac{R_1}{R_0^2(\beta - \log R_0)} + \frac{R_0[1 + 2\beta - 2\log R_0]}{4R_0^2(\beta - \log R_0)^3} = -\dot{R}_1,$$

which eventually yields the following first-order differential equation for R_1 as a function of R_0, namely

$$R_0(\beta - \log R_0)\frac{dR_1}{dR_0} + (\beta - 1 - \log R_0)R_1$$

$$= -\frac{1 + 2\beta + 2\beta^2}{4R_0(\beta - \log R_0)^2}$$

$$+ \frac{R_0[2(\log R_0)^2 - 2(1 + 2\beta)\log R_0 + (1 + 2\beta + 2\beta^2)]}{4(\beta - \log R_0)^2}. \qquad (2.58)$$

On solving this equation using R_1 zero when R_0 is unity we readily obtain

$$R_1(\bar{t}) = \frac{1 + 2\beta + R_0^2}{4R_0(\beta - \log R_0)} - \frac{1 + 2\beta + 2\beta^2 - R_0^2}{4R_0(\beta - \log R_0)^2}. \qquad (2.59)$$

From

$$R_0 = R - \frac{\{(1 + 2\beta + R^2) - (1 + 2\beta + 2\beta^2 - R^2)/(\beta - \log R)\}}{4\alpha R(\beta - \log R)} + O\!\left(\frac{1}{\alpha^2}\right),$$

$$(2.60)$$

(2.26) and (2.56) we may deduce the order one corrected motion

$$t = \frac{\alpha}{4}[2R^2 \log R + (1+2\beta)(1-R^2)]$$

$$+ \frac{1}{4}\left\{(1+2\beta+R^2) - \frac{1+2\beta+2\beta^2-R^2}{(\beta-\log R)}\right\} + O\left(\frac{1}{\alpha}\right), \qquad (2.61)$$

and again we remark that although (2.59) and (2.60) give singular results as the boundary approaches the centre of the cylinder we observe that (2.61) remains finite in this limit. We may verify that the final expression for the temperature up to order α^{-1} becomes

$$T(r, t) = (\log r - \log R_0)/(\beta - \log R_0)$$

$$+ \left\{(1+2\beta+2\beta^2-R_0^2)(\beta-\log r) - (1+2\beta+2\beta^2-r^2)(\beta-\log R_0)\right.$$

$$+ \left[(1+2\beta+2R_0^2-r^2)\log R_0 + 1 + \beta + \beta r^2 - (1+2\beta)R_0^2](\beta-\log r)\right\} \Big/$$

$$4\alpha R_0^2(\beta-\log R_0)^4 + O\left(\frac{1}{\alpha^2}\right). \qquad (2.62)$$

Again we observe that we may readily check that the given expression satisfies the boundary conditions (2.5) up to order α^{-1}. Moreover, for β zero the above results coincide with those given by Pekeris and Slichter (1939). In the following sections we deduce the results given here by a more direct process.

2.3 Freezing an infinite slab

Following Huang and Shih (1975a) we present in this section a perturbation solution for a large Stefan number for planar solidification of a saturated liquid with convection at the wall. In contrast to the previous section we derive these solutions by, firstly, fixing the boundary by Landau's transformation (2.8) and, secondly, utilizing the boundary position $X(t)$ as the time variable. This enables the Stefan phase change condition (2.3)$_1$ to be actually incorporated in the partial differential equation for the temperature. Thus we introduce the transformations

$$\rho = \frac{x}{X(t)}, \qquad \tau = X(t), \qquad T(x, t) = u(\rho, \tau), \qquad (2.63)$$

so that (2.1) and (2.2) become, respectively,

$$\alpha u_{\rho\rho} = u_\rho(1, \tau)(\rho u_\rho - \tau u_\tau), \qquad (2.64)$$

$$\beta u_\rho(0, \tau) = \tau[u(0, \tau) - 1], \qquad u(1, \tau) = 0, \qquad (2.65)$$

where subscripts denote partial derivatives and the arguments of u and its partial derivatives are understood to be (ρ, τ) unless otherwise indicated. In these variables $(2.3)_1$ becomes

$$u_\rho(1, \tau) = -\alpha\tau \frac{d\tau}{dt}, \qquad \tau(0) = 0. \tag{2.66}$$

We now solve (2.64) and (2.65) for large α by a regular perturbation series,

$$u(\rho, \tau) = u_0(\rho, \tau) + \frac{u_1(\rho, \tau)}{\alpha} + \frac{u_2(\rho, \tau)}{\alpha^2} + O\left(\frac{1}{\alpha^3}\right), \tag{2.67}$$

and obtain the corresponding approximation to the motion of the boundary by substituting (2.67) into (2.66) and integrating

$$dt = -\alpha\tau \left\{ u_{0\rho}(1, \tau) + \frac{u_{1\rho}(1, \tau)}{\alpha} + \frac{u_{2\rho}(1, \tau)}{\alpha^2} + \ldots \right\}^{-1} d\tau, \tag{2.68}$$

by expanding the integrand in powers of α^{-1}. In Chapter 6 we present an alternative approximate solution of (2.64) and (2.65) which relates to the classical Neumann solution. The choice of variables (2.63) means that for both approximate solutions there are essentially two stages to the procedure. Firstly, we solve (2.64) and (2.65) to obtain an estimate of the temperature and then we use this estimate in (2.66) to obtain an expression for the motion of the boundary. It is important to appreciate that in performing this integration further approximations are made simply to obtain analytic expressions. These later approximations, however, such as the expansion of the integrand in (2.68) in powers of α^{-1}, are of course consistent with the approximations employed in solving (2.64) and (2.65).

On substituting (2.67) into (2.64) and (2.65) we obtain the following boundary-value problems for the first three terms,

$$u_{0\rho\rho} = 0, \tag{2.69}$$
$$\beta u_{0\rho}(0, \tau) = \tau[u_0(0, \tau) - 1], \qquad u_0(1, \rho) = 0,$$

$$u_{1\rho\rho} = u_{0\rho}(1, \tau)(\rho u_{0\rho} - \tau u_{0\tau}), \tag{2.70}$$
$$\beta u_{1\rho}(0, \tau) = \tau u_1(0, \tau), \qquad u_1(1, \rho) = 0,$$

and

$$u_{2\rho\rho} = u_{1\rho}(1, \tau)(\rho u_{0\rho} - \tau u_{0\tau}) + u_{0\rho}(1, \tau)(\rho u_{1\rho} - \tau u_{1\tau}), \tag{2.71}$$
$$\beta u_{2\rho}(0, \tau) = \tau u_2(0, \tau), \qquad u_2(1, \rho) = 0.$$

After a long but straightforward calculation we obtain

$$u_0(\rho, \tau) = \frac{\tau(1-\rho)}{(\tau+\beta)}, \tag{2.72}$$

$$u_1(\rho, \tau) = -\frac{\tau^2(1-\rho)}{6(\tau+\beta)^4}\left\{\tau(\tau+\beta)\rho^2 + (\tau+\beta)(\tau+3\beta)\rho + (\tau+3\beta)\beta\right\}, \tag{2.73}$$

$$u_2(\rho, \tau) = \frac{\tau^3(1-\rho)}{360(\tau+\beta)^7}\left\{9\tau^2(\tau+\beta)^2\rho^4 + 9\tau(\tau+\beta)^2(\tau+5\beta)\rho^3 \right.$$
$$+ \tau(\tau+\beta)(19\tau^2 + 84\tau\beta + 165\beta^2)\rho^2$$
$$\left. + (19\tau^3 + 114\tau^2\beta + 255\tau\beta^2 + 360\beta^3)[(\tau+\beta)\rho + \beta]\right\}. \tag{2.74}$$

We note that the formula given by Huang and Shih (1975a) corresponding to (2.74) contains a minor typing error. From the above expressions and (2.68) we obtain

$$t = \frac{\alpha}{2}\tau(\tau+2\beta) + \frac{\tau^2(\tau+3\beta)}{6(\tau+\beta)} - \frac{\tau^3(\tau^3 + 6\tau^2\beta + 15\tau\beta^2 + 15\beta^3)}{45\alpha(\tau+\beta)^4} + O\left(\frac{1}{\alpha^2}\right). \tag{2.75}$$

Thus, altogether in the original variables we have for the temperature

$$T(x, t) = \left(\frac{X-x}{X+\beta}\right)\left[1 - \frac{(X+\beta)x^2 + (X+\beta)(X+3\beta)x + (X+3\beta)X\beta}{6\alpha(X+\beta)^3}\right.$$
$$+ \left\{\begin{array}{l}9(X+\beta)^2x^4 + 9(X+\beta)^2(X+5\beta)x^3 \\ + X(X+\beta)(19X^2 + 84X\beta + 165\beta^2)x^2 + X(19X^3 \\ + 114X^2\beta + 255X\beta^2 + 360\beta^3)[(X+\beta)x + X\beta]\end{array}\right\} \Bigg/$$
$$\left. 360\alpha^2(X+\beta)^6\right] + O\left(\frac{1}{\alpha^3}\right), \tag{2.76}$$

while the motion of the boundary $X(t)$ is obtained from

$$t = \frac{\alpha}{2}X(X+2\beta) + \frac{X^2(X+3\beta)}{6(X+\beta)} - \frac{X^3(X^3 + 6X^2\beta + 15X\beta^2 + 15\beta^3)}{45\alpha(X+\beta)^4}$$
$$+ O\left(\frac{1}{\alpha^2}\right). \tag{2.77}$$

We observe that up to order α^{-1} we may reconcile (2.76) with (2.37) and (2.40) of the previous section. These results only apply for large α, since from (2.77) it is apparent that if α is sufficiently small then times may become negative. For β zero the above expressions coincide with those given by Huang and Shih (1975c), who derive these expressions directly from Neumann's solution and from the above perturbation scheme.

We remark that Pedroso and Domoto (1973a) also formulate the above perturbation solution but in such a way that the algebra can be effected on a computer. These authors give numerical values of the coefficients $\bar{t}_i(X/\beta)$ for $i = 1$–9 in the series

$$\frac{t}{\alpha\beta^2} = \sum_{i=1}^{9} \bar{t}_i \left(\frac{X}{\beta}\right)\left(\frac{1}{\alpha}\right)^{i-1}, \tag{2.78}$$

for twelve values of X/β, the first three terms of (2.77) giving rise to exactly their numerical values for $\bar{t}_i(X/\beta)$ for $i = 1$, 2 and 3. Their series is convergent for $\alpha \geq 1$. For example, for $\alpha = 2$ and $X/\beta = 1$ the first four terms of (2.78) give $t/\alpha\beta^2 = 1.656$ and this value is unaffected by the addition of higher terms. In the following chapter we show that the first term of (2.77) and the first two terms of (2.77) constitute lower and upper bounds, respectively, to the actual motion of the boundary $t(X)$. Accordingly in Tables 2.1 and 2.2 we give numerical values of the various estimates of $t/\alpha\beta^2$ for $\alpha = 1$ and 2 and for the twelve values of X/β used by Pedroso and Domoto (1973a). The first estimate is the lower bound, the second is based on three terms of (2.77), the third is calculated from the nine coefficients given by Pedroso and Domoto (1973a) and the fourth estimate is the upper bound.

Finally in this section we note two extensions of the analysis presented

Table 2.1 Values of various estimates of $t/\alpha\beta^2$ for $\alpha = 1$ and twelve values of X/β used by Pedroso and Domoto (1973a)

X/β	$t_{pss}(X)$	Three terms (2.77)	Nine terms (2.78)	Two terms (2.77)
0.2	0.2200	0.2362	0.2365	0.2378
0.4	0.4800	0.5366	0.5385	0.5448
0.6	0.7800	0.8957	0.9005	0.9150
0.8	1.1200	1.3112	1.3199	1.3452
1.0	1.5000	1.7820	1.7953	1.8333
1.4	2.3800	2.8861	2.9103	2.9789
1.8	3.4200	4.2039	4.2405	4.3457
2.2	4.6200	5.7330	5.7835	5.9308
2.6	5.9800	7.4718	7.5376	7.7326
3.0	7.5000	9.4195	9.5015	9.7500
4.0	12.0000	15.1986	15.3272	15.7300
5.0	17.5000	22.2732	22.4575	23.0556

Table 2.2 Values of various estimates of $t/\alpha\beta^2$ for $\alpha = 2$ and twelve values of X/β used by Pedroso and Domoto (1973a)

X/β	$t_{pss}(X)$	Three terms (2.77)	Nine terms (2.78)	Two terms (2.77)
0.2	0.2200	0.2285	0.2285	0.2289
0.4	0.4800	0.5103	0.5106	0.5124
0.6	0.7800	0.8427	0.8434	0.8475
0.8	1.1200	1.2241	1.2255	1.2326
1.0	1.5000	1.6538	1.6559	1.6667
1.4	2.3800	2.6562	2.6600	2.6794
1.8	3.4200	3.8474	3.8531	3.8829
2.2	4.6200	5.2260	5.2339	5.2754
2.6	5.9800	6.7911	6.8016	6.8563
3.0	7.5000	8.5424	8.5554	8.6250
4.0	12.0000	13.7330	13.7537	13.8667
5.0	17.5000	20.0822	20.1123	20.2778

here. Weinbaum and Jiji (1977) use singular perturbation theory for the problem of melting or freezing in a finite domain which is initially not at the fusion temperature. One end of a finite slab is maintained at constant temperature while the opposite face is assumed to be either insulated or kept at the initial temperature. Perturbation solutions for a large Stefan number are obtained for both the short time scale characterizing the transient diffusion in the liquid phase and the long time scale characterizing the interface motion. Again for large α, Yan and Huang (1979) obtain perturbation solutions for the one-dimensional phase-change problem in a finite region subject to a convective and radiative boundary condition at the fixed boundary. As usual the radiative term is approximated by a Taylor series expansion. Analytical expressions are obtained for the temperature and moving boundary and numerical values are in close agreement with Chung and Yeh (1975).

2.4 Freezing a liquid in a spherical container

In this and the following section we describe the formal derivation of large α expansions for the freezing of spherical and infinite cylindrical containers of liquids initially at the fusion temperature. The derivations follow Huang and Shih (1975b), although Pedroso and Domoto (1973b) were first to obtain regular perturbation solutions for spherical solidification but with no Newton cooling on the surface. Both solutions contain singularities as the solidification front approaches the centre of either the

sphere or cylinder. However, we emphasize that the order one corrections to the motion as given by (2.51) and (2.61) are not singular and for
our purposes these are the important results, since we show in the
following chapter that they constitute an upper bound to the motion. In
the final section of this chapter we give a brief review of some of the
developments attempting to overcome the singular nature of the
expansions.

For spherical solidification we introduce new variables by means of the
transformations

$$\rho = \frac{1-r}{1-R}, \qquad \tau = 1 - R, \qquad T(r, t) = \frac{u(\rho, \tau)}{r}, \tag{2.79}$$

so that (2.4) and (2.5) become

$$\alpha(1-\tau)u_{\rho\rho} = u_\rho(1, \tau)(\rho u_\rho - \tau u_\tau), \tag{2.80}$$

$$\beta u_\rho(0, \tau) = \tau[(1-\beta)u(0, \tau) - 1], \qquad u(1, \tau) = 0, \tag{2.81}$$

while from (2.6) we obtain

$$u_\rho(1, \tau) = -\alpha\tau(1-\tau)\frac{d\tau}{dt}, \qquad \tau(0) = 0. \tag{2.82}$$

On looking for a solution of the form (2.67) for large α we obtain the
following boundary-value problems for the first three terms:

$$u_{0\rho\rho} = 0, \tag{2.83}$$

$$\beta u_{0\rho}(0, \tau) = \tau[(1-\beta)u_0(0, \tau) - 1], \qquad u_0(1, \tau) = 0,$$

$$(1-\tau)u_{1\rho\rho} = u_{0\rho}(1, \tau)(\rho u_{0\rho} - \tau u_{0\tau}),$$
$$\beta u_{1\rho}(0, \tau) = \tau(1-\beta)u_1(0, \tau), \qquad u_1(1, \tau) = 0, \tag{2.84}$$

and

$$(1-\tau)u_{2\rho\rho} = u_{1\rho}(1, \tau)(\rho u_{0\rho} - \tau u_{0\tau}) + u_{0\rho}(1, \tau)(\rho u_{1\rho} - \tau u_{1\tau}), \tag{2.85}$$
$$\beta u_{2\rho}(0, \tau) = \tau(1-\beta)u_2(0, \tau), \qquad u_2(1, \tau) = 0.$$

After a long calculation we obtain

$$u_0(\rho, \tau) = \frac{\tau(1-\rho)}{[\beta + (1-\beta)\tau]}, \tag{2.86}$$

$$u_1(\rho, \tau) = \frac{\tau^2}{6(1-\tau)[\beta + (1-\beta)\tau]^4}\Big\{[\beta + (1-\beta)\tau][3\beta + (1-\beta)\rho\tau]\rho^2$$
$$- [\beta + (1-\beta)\rho\tau][3\beta + (1-\beta)\tau]\Big\}, \tag{2.87}$$

$$u_2(\rho, \tau) =$$

$$\frac{\tau^3}{360(1-\tau)^3[\beta + (1-\beta)\tau]^7}\left\{3\tau[\beta + (1-\beta)\tau][4\beta - 3 + 4(1-\beta)\tau]\right.$$

$$\times \{[\beta + (1-\beta)\tau][5\beta + (1-\beta)\rho\tau]\rho^4 - [\beta + (1-\beta)\rho\tau][5\beta + (1-\beta)\tau]\}$$

$$+ 10[(\beta - 1)\tau^2 + 3\beta(4\beta - 1)\tau - 12\beta^2]$$

$$\left.\times \{[\beta + (1-\beta)\tau][3\beta + (1-\beta)\rho\tau]\rho^2 - [\beta + (1-\beta)\rho\tau][3\beta + (1-\beta)\tau]\}\right\}.$$

In the original variables these results become   (2.88)

$$u_0 = \frac{r - R}{[1 + (\beta - 1)R]}, \tag{2.89}$$

$$u_1 = \frac{1}{6R[1 + (\beta - 1)R]^4}\left\{[1 + (\beta - 1)R][1 + 2\beta + (\beta - 1)r](1-r)^2\right.$$

$$\left. - [1 + (\beta - 1)r][1 + 2\beta + (\beta - 1)R](1-R)^2\right\}, \tag{2.90}$$

$$u_2 = \frac{1}{360R^3[1 + (\beta - 1)R]^7}\left\{3[1 + (\beta - 1)R][1 + 4(\beta - 1)R]\right.$$

$$\times \{[1 + (\beta - 1)R][1 + 4\beta + (\beta - 1)r](1-r)^4$$

$$- [1 + (\beta - 1)r][1 + 4\beta + (\beta - 1)R](1-R)^4\}$$

$$+ 10(1-R)[(\beta - 1)R^2 - R(12\beta^2 - \beta - 2) - (1 + 2\beta)]$$

$$\times \{[1 + (\beta - 1)R][1 + 2\beta + (\beta - 1)r](1-r)^2$$

$$\left. - [1 + (\beta - 1)r][1 + 2\beta + (\beta - 1)R](1-R)^2\}\right\}. \tag{2.91}$$

On integrating

$$dt = -\alpha\tau(1-\tau)\left\{u_{0\rho}(1, \tau) + \frac{u_{1\rho}(1, \tau)}{\alpha} + \frac{u_{2\rho}(1, \tau)}{\alpha^2} + \ldots\right\}^{-1} d\tau, \tag{2.92}$$

by expanding the integrand in powers of α^{-1} we obtain from the above formulae, after a long calculation, the following expression for the motion of the boundary:

$$t = \frac{\alpha}{6}(1-R)[(1 + 2\beta)(1 + R) + 2(\beta - 1)R^2]$$

$$+ \frac{(1-R)^2[1 + 2\beta + (\beta - 1)R]}{6[1 + (\beta - 1)R]}$$

$$- \frac{(1-R)^3\left\{\begin{array}{c}15\beta^3 + 15\beta^2(1-\beta)(1-R) + 16\beta(1-\beta)^2 \\ \times (1-R)^2 + (1-\beta)^3(1-R)^3\end{array}\right\}}{45\alpha R[1 + (\beta - 1)R]^4}$$

$$+ O\left(\frac{1}{\alpha^2}\right). \tag{2.93}$$

We again observe that the pseudo-steady-state emerges as the leading term, that the first two terms are consistent with the results given in Section 2 of this chapter and that both u_1 and u_2, but only the order α^{-1} term and higher in (2.93), are singular as the boundary approaches the centre of the sphere. We remark that in this and the following section the corresponding formulae given by Huang and Shih (1975b) contain several typing errors.

2.5 Freezing an infinite circular cylinder

For cylindrical freezing we introduce new variables given by

$$\rho = \frac{\log r}{\log R}, \qquad \tau = \log R, \qquad T(r, t) = u(\rho, \tau), \tag{2.94}$$

so that (2.5), (2.6) and (2.7) yield

$$\alpha\, e^{2\tau(1-\rho)} u_{\rho\rho} = u_\rho(1, \tau)(\rho u_\rho - \tau u_\tau), \tag{2.95}$$

$$\beta u_\rho(0, \tau) = \tau[1 - u(0, \tau)], \qquad u(1, \tau) = 0, \tag{2.96}$$

$$u_\rho(1, \tau) = -\alpha\, e^{2\tau}\tau \frac{d\tau}{dt}, \qquad \tau(0) = 0. \tag{2.97}$$

Again, assuming an expansion of the form (2.67), we may deduce the following boundary-value problems for the first three terms:

$$u_{0\rho\rho} = 0,$$
$$\beta u_{0\rho}(0, \tau) = \tau[1 - u_0(0, \tau)], \qquad u_0(1, \tau) = 0, \tag{2.98}$$

$$e^{2\tau(1-\rho)} u_{1\rho\rho} = u_{0\rho}(1, \tau)(\rho u_{0\rho} - \tau u_{0\tau}),$$
$$\beta u_{1\rho}(0, \tau) = -\tau u_1(0, \tau), \qquad u_1(1, \tau) = 0, \tag{2.99}$$

and

$$e^{2\tau(1-\rho)} u_{2\rho\rho} = u_{1\rho}(1, \tau)(\rho u_{0\rho} - \tau u_{0\tau}) + u_{0\rho}(1, \tau)(\rho u_{1\rho} - \tau u_{1\tau}), \tag{2.100}$$

$$\beta u_{2\rho}(0, \tau) = -\tau u_2(0, \tau), \qquad u_2(1, \tau) = 0.$$

On solving these problems we obtain

$$u_0(\rho, \tau) = \frac{\tau(\rho - 1)}{(\beta - \tau)}, \tag{2.101}$$

$$u_1(\rho, \tau) = \frac{e^{-2\tau}}{4(\beta - \tau)^4}\Big\{(\beta - \tau)(1 + \beta - \rho\tau)\, e^{2\rho\tau} - (\beta - \rho\tau)$$
$$\times (1 + \beta - \tau)\, e^{2\tau} - (1 + 2\beta + 2\beta^2)\tau(\rho - 1)\Big\}, \tag{2.102}$$

$$u_2(\rho, \tau) = \frac{e^{-4\tau}}{128(\beta - \tau)^7}\Big\{(\beta - \tau)(2\beta - 2\tau - 3)[(\beta - \tau)(2\beta + 3 - 2\rho\tau)\,e^{4\rho\tau}$$
$$- (\beta - \rho\tau)(2\beta + 3 - 2\tau)\,e^{4\tau} - (3 + 12\beta + 8\beta^2)\tau(\rho - 1)]$$
$$+ 8[5 + 5(\beta - \tau) + 2(\beta - \tau)^2]\,e^{2\tau}[(\beta - \tau)(\beta + 1 - \rho\tau)\,e^{2\rho\tau}$$
$$- (\beta - \rho\tau)(\beta + 1 - \tau)\,e^{2\tau} - (1 + 2\beta + 2\beta^2)\tau(\rho - 1)]$$
$$- 40(1 + 2\beta + 2\beta^2)[(\beta - \tau)\,e^{2\rho\tau} - (\beta - \rho\tau)\,e^{2\tau}$$
$$- (1 + 2\beta)\tau(\rho - 1)] - 8(1 + 2\beta + 2\beta^2)(2\beta - 2\tau - 5)$$
$$\times \tau(\rho - 1)[(\beta - \tau)\,e^{2\rho\tau} + (1 + 2\beta)\tau - \beta]\Big\}. \qquad (2.103)$$

In terms of the original variables these results become

$$u_0 = \frac{\log r - \log R}{\beta - \log R}, \qquad (2.104)$$

$$u_1 = \frac{1}{4R^2(\beta - \log R)^4}\Big\{(\beta - \log R)(\beta + 1 - \log r)r^2 - (\beta - \log r)$$
$$\times (\beta + 1 - \log R)R^2 - (1 + 2\beta + 2\beta^2)(\log r - \log R)\Big\}, \qquad (2.105)$$

$$u_2 = \frac{1}{128R^4(\beta - \log R)^7}\Big\{(\beta - \log R)(2\beta - 3 - 2\log R)[(\beta - \log R)$$
$$\times (2\beta + 3 - 2\log r)r^4 - (\beta - \log r)(2\beta + 3 - 2\log R)R^4$$
$$- (3 + 12\beta + 8\beta^2)(\log r - \log R)] + 8[5 + 5(\beta - \log R)$$
$$+ 2(\beta - \log R)^2]R^2[(\beta - \log R)(\beta + 1 - \log r)r^2 - (\beta - \log r)$$
$$\times (\beta + 1 - \log R)R^2 - (1 + 2\beta + 2\beta^2)(\log r - \log R)]$$
$$- 40(1 + 2\beta + 2\beta^2)[(\beta - \log R)r^2 - (\beta - \log r)R^2 - (1 + 2\beta)$$
$$\times (\log r - \log R)] - 8(1 + 2\beta + 2\beta^2)(2\beta - 5 - 2\log R)$$
$$\times (\log r - \log R)[(\beta - \log R)r^2 + (1 + 2\beta)\log R - \beta]\Big\}, \qquad (2.106)$$

and again we observe the singular nature of u_1 and u_2 as the boundary approaches the centre of the cylinder. From these results and integrating (2.97) we obtain for the motion of the boundary

$$t = \frac{\alpha}{4}[2R^2 \log R + (1 + 2\beta)(1 - R^2)]$$

$$+ \frac{1}{4}\Big\{(1 + 2\beta + R^2) - \frac{1 + 2\beta + 2\beta^2 - R^2}{(\beta - \log R)}\Big\}$$

$$+ \frac{1}{128\alpha R^2(\beta - \log R)^4}\Big\{[8(\beta - \log R)^3 + 20(\beta - \log R)^2$$

$$+ 21(\beta - \log R) + 8]R^4 - 8(1 + 2\beta + 2\beta^2)[2(\beta + 1 - \log R)R^2$$
$$- (1 + 2\beta + 2\beta^2)] - (5 + 20\beta + 40\beta^2 + 32\beta^3)(\beta - \log R)\big\}$$
$$+ O\left(\frac{1}{\alpha^2}\right). \tag{2.107}$$

The solutions presented here and in the previous section are partial in the sense that they do not apply for times close to complete solidification. The various attempts to accommodate the singular nature of these results are reviewed in the following section.

2.6 Review of subsequent developments

Pedroso and Domoto (1973b, 1973c) attempt to improve the convergence of the given series by means of various *ad hoc* techniques, while Riley *et al.* (1974), Stewartson and Waechter (1976) and Soward (1980) present a large α matched asymptotic expansion theory. The latter approach altogether providing a good deal of insight into the essential nature of the freezing process near the centre does, however, involve considerable mathematical analysis. Moreover, although the singularity as $R(t)$ approaches zero is accommodated, the results obtained and the physical picture presented apply only for a large Stefan number. For these reasons, and the fact that these authors only consider the case of β zero, we present only a review of these developments and for the most part we confine our attention to spherical freezing.

For β zero (2.89)–(2.91) simplify to give

$$u_0 = \left(\frac{r - R}{1 - R}\right), \tag{2.108}$$

$$u_1 = \frac{1}{6R}\left\{\left(\frac{1-r}{1-R}\right)^3 - \left(\frac{1-r}{1-R}\right)\right\}, \tag{2.109}$$

$$u_2 = \frac{1}{360R^3}\left\{3(1 - 4R)\left[\left(\frac{1-r}{1-R}\right)^5 - \left(\frac{1-r}{1-R}\right)\right]\right.$$
$$\left. - 10\left[\left(\frac{1-r}{1-R}\right)^3 - \left(\frac{1-r}{1-R}\right)\right]\right\}, \tag{2.110}$$

while from (2.93) the motion of the boundary becomes

$$t = \frac{\alpha}{6}(1 - R)^2(1 + 2R) + \frac{(1 - R)^2}{6} - \frac{(1 - R)^2}{45\alpha R} + O\left(\frac{1}{\alpha^2}\right). \tag{2.111}$$

Although not relevant to inward solidification we note that Pedroso and

Domoto (1973b) improve the convergence of the corresponding series for outward solidification by utilizing the non-linear transformations due to Shanks (1955), which apply when the perturbation series is either slowly convergent or divergent. For example, if three terms of the series are known,

$$T(r, t) = \frac{1}{r}\left\{u_0 + \frac{u_1}{\alpha} + \frac{u_2}{\alpha^2} + O\left(\frac{1}{\alpha^3}\right)\right\}, \tag{2.112}$$

then the simplest of the transformations due to Shanks (1955) gives

$$T(r, t) \simeq \frac{\{u_0 + (u_1 - u_0 u_2/u_1)/\alpha\}}{r(1 - u_2/\alpha u_1)}, \tag{2.113}$$

which on expansion in powers of α^{-1} agrees with (2.112). Unfortunately, for inward solidification this simple transformation does not improve the singularity as $R(t)$ tends to zero. For inward solidification Pedroso and Domoto (1974b) expand the temperature in powers of R^{-1}, thus

$$T(r, t) = \frac{1}{r}\left\{u_0 + \frac{v_1}{\alpha R} + \frac{v_3}{\alpha^2 R^3} + \ldots\right\}, \tag{2.114}$$

where v_1 and v_3 are finite as R tends to zero and are readily identified from (2.109) and (2.110), respectively. These authors then adopt an Euler transformation (see, for example, Meksyn 1961, p. 58) with respect to R^{-1}, namely

$$S = \frac{R^{-1}}{K(\alpha) + R^{-1}} = \frac{1}{1 + K(\alpha)R}, \tag{2.115}$$

so that S tends to unity as R approaches zero and the parameter $K(\alpha)$ is determined from an overall energy balance from the instant freezing begins to the moment of complete solidification. By integrating (2.4) with respect to r from $R(t)$ to 1 we may deduce

$$\frac{d}{dt}\left(\int_{R(t)}^1 r^2 T(r, t)\, dr\right) = \frac{\partial T}{\partial r}(1, t) + \alpha R^2 \frac{dR}{dt}, \tag{2.116}$$

and a further integration with respect to time and setting R zero yields the overall energy balance equation,

$$\int_0^1 \frac{\partial T^\dagger}{\partial r}(1, R)\left\{\frac{\partial T^\dagger}{\partial r}(R, R)\right\}^{-1} dR - \frac{1}{\alpha}\int_0^1 r^2 T^\dagger(r, 0)\, dr = \frac{1}{3}, \tag{2.117}$$

where $T^\dagger(r, R)$ denotes the temperature $T(r, t)$ but employing (r, R) as independent variables. From (2.114) and (2.115) we have

$$T^\dagger(r, R) = \frac{1}{r}\left\{u_0 + \frac{v_1}{\alpha} KS(1 + S) + \frac{KS^3}{\alpha^2}(v_3 K^2 + \alpha v_1) + \ldots\right\}, \tag{2.118}$$

and truncating this series at the ith power of S, where i is 1, 2 or 3, the energy balance equation (2.117) may be used to obtain a transcendental equation for $K_1(\alpha)$, $K_2(\alpha)$ or $K_3(\alpha)$, which hopefully provide a convergent sequence to $K(\alpha)$. These transcendental equations rapidly become complicated and we refer the reader to Pedroso and Domoto (1973b) for further details. We remark, however, that their numerical results appear to indicate that the estimates $K_i(\alpha)$ do indeed converge.

Pedroso and Domoto (1973c) using r and R as independent variables look for transformations

$$\phi = F(r, R, \alpha), \qquad \Phi = G(R, \alpha), \tag{2.119}$$

where $G(R, \alpha) = F(R, R, \alpha)$, such that the temperature $T^\dagger(r, R)$ is given exactly by the pseudo-steady-state form, namely

$$T = \frac{1 - \Phi/\phi}{1 - \Phi}. \tag{2.120}$$

The inverse of (2.119) are assumed to take the form

$$r = \phi + \frac{f_1(\phi, \Phi)}{\alpha} + \frac{f_2(\phi, \Phi)}{\alpha^2} + O\left(\frac{1}{\alpha^3}\right),$$

$$R = \Phi + \frac{f_1(\Phi, \Phi)}{\alpha} + \frac{f_2(\Phi, \Phi)}{\alpha^2} + O\left(\frac{1}{\alpha^3}\right), \tag{2.121}$$

where

$$f_1(1, \Phi) = f_2(1, \Phi) = 0, \tag{2.122}$$

$$\lim_{\Phi \to 1} f_1(\Phi, \Phi) = \lim_{\Phi \to 1} f_2(\Phi, \Phi) = 0.$$

On transforming the basic partial differential equation

$$\frac{\partial^2 T^\dagger}{\partial r^2} + \frac{2}{r} \frac{\partial T^\dagger}{\partial r} + \frac{1}{\alpha} \frac{\partial T^\dagger}{\partial R} \frac{\partial T^\dagger}{\partial r} (R, R) = 0, \tag{2.123}$$

with (ϕ, Φ) as independent variables in place of (r, R), Pedroso and Domoto (1973c) show that for $T^\dagger$ given by (2.120) the functions f_1 and f_2 are found to be given by

$$f_1(\phi, \Phi) = -\frac{\phi(1 - \phi)^3}{6\Phi^2(1 - \Phi)^2}, \tag{2.124}$$

$$f_2(\phi, \Phi) = \frac{[40\Phi(1 - \Phi)^2 - 3(1 - 4\Phi + 10\phi)(1 - \phi)^2]\phi(1 - \phi)^3}{360\Phi^4(1 - \Phi)^4}. \tag{2.125}$$

In particular, the motion of the boundary is determined parametrically from the equations

$$R = \Phi - \frac{1 - \Phi}{6\alpha\Phi} + \frac{(22\Phi - 3)(1 - \Phi)}{360\alpha^2\Phi^3} + O\left(\frac{1}{\alpha^3}\right), \tag{2.126}$$

$$t = \frac{\alpha}{6}(1 - \Phi)^2(1 + 2\Phi) + \frac{(1 - \Phi)^2}{3} - \frac{(1 - \Phi)^2}{180\alpha\Phi^2} + O\left(\frac{1}{\alpha^2}\right). \tag{2.127}$$

Unfortunately, numerical results based on (2.126) and (2.127) do not agree with Tao (1967), especially for large α, where of course the solution should be accurate, and we refer the reader to Pedroso and Domoto (1973c) for the full discussion. We remark, however, that on inverting (2.126) we may deduce

$$\Phi = R + \frac{1 - R}{6\alpha R} - \frac{(22R + 7)(1 - R)}{360\alpha^2R^3} + O\left(\frac{1}{\alpha^3}\right), \tag{2.128}$$

which on substitution into (2.127) yields exactly (2.111), so that perhaps this procedure does not after all accommodate the singularity as $R(t)$ approaches the centre.

Riley et al. (1974) were the first authors to attempt to accommodate the singularity by means of the method of matched asymptotic expansions. Subsequently, Stewartson and Waechter (1976) showed that even the Riley et al. (1974) solution has a singular behaviour near the centre for times close to complete freezing and consequently determined a new asymptotic solution. Soward (1980) presented a compact account of these developments and rederived a number of earlier results by simpler methods. The following account is based on this latter work for which it is convenient to employ the starred physical variables introduced in Chapter 1 (see (1.25)–(1.29)).

The problems of freezing infinite circular cylinders or spheres are characterized by two distinct time scales. The first is the time taken for heat to diffuse a distance a, which is of order

$$t_D^* = \frac{a^2\rho_1C_1}{k_1}, \tag{2.129}$$

while the second is the time taken for complete freezing

$$t_F^* = \frac{a^2\rho_1L}{k_1(T_f - T_0)}\tau_e, \tag{2.130}$$

where the magnitude of the constant τ_e depends on the Stefan number $\alpha = L/C_1(T_f - T_0)$. In the limit of large α, the time scales separate

$$t_F^* \gg t_D^*, \tag{2.131}$$

and the expansions presented in this chapter show that τ_e is of order one ($\frac{1}{6}$ for the sphere and $\frac{1}{4}$ for the cylinder). When

$$t^* \ll t_F^* - t_D^* \tag{2.132}$$

the boundary moves slowly and, measured on the time scale t_F^*, the temperature distribution in the solid responds rapidly to the continual change of the solid–liquid interface and the time derivative in the heat-conduction equation may be neglected. On the other hand, for times close to complete solidification,

$$t^* = t_F^* + O(t_D^*), \tag{2.133}$$

the large α expansions are singular and no approximation can be made in the heat-conduction equation, so that a new asymptotic solution is appropriate. This solution was first obtained by Riley et al. (1974). These authors obtained the value of τ_e correct to order $\alpha^{-3/2}$ for the sphere and correct to order $\alpha^{-1} (\log \Lambda)^{-2}$ for the cylinder, where Λ is determined from

$$\alpha = \frac{\Lambda^2}{\log \Lambda^2}, \tag{2.134}$$

which, incidentally, exists only if $\alpha \ge e$. These results are important and worth while summarizing.

For spherical freezing with no Newton cooling ($\beta = 0$) we have

$$\tau_e = \tau_s + \frac{1}{6} + \frac{1}{6\alpha} + O\left(\frac{1}{\alpha^2}\right), \tag{2.135}$$

where

$$\tau_s = -\frac{2^{1/2}\zeta(3)}{\alpha^{3/2}\pi^{5/2}} + O\left(\frac{1}{\alpha^2}\right), \tag{2.136}$$

where $\zeta(3) = 1.2020569\ldots$. Further, Riley et al. (1974) obtain the following approximate solution for the temperature valid for times close to complete freezing and positions $r \gg (\alpha\tau_e - t)^{1/2}$,

$$T(r, t) = 1 + \frac{2^{1/2}}{\alpha^{1/2}\pi^{3/2}r} \int_0^{\pi r} \log\left(2 \sin\frac{\xi}{2}\right) d\xi + O\left(\frac{1}{\alpha}\right). \tag{2.137}$$

For cylindrical freezing with no Newton cooling these authors obtain

$$\tau_e = \tau_c + \frac{1}{4} + \frac{1}{4\alpha} + O\left(\frac{1}{\alpha^2}\right), \tag{2.138}$$

where

$$\tau_c = \frac{1}{\alpha}\left\{-\frac{(\log \Lambda)^{-1}}{4} + (\log \Lambda)^{-2} \sum_{n=1}^{\infty} \frac{\gamma + 2\log(p_n/2)}{p_n^4 J_1(p_n)^2}\right.$$

$$\left. + O((\log \Lambda)^{-3})\right\}, \tag{2.139}$$

where here $\gamma = 0.5772157\ldots$ is Euler's constant and p_n is the nth positive zero of the zero order Bessel function $J_0(p)$. Moreover, for times close to complete freezing and positions $r \gg (\alpha\tau_e - t)^{1/2}$ we have the approximate solution

$$T(r, t) = 1 + (\log \Lambda)^{-1} \log r$$

$$+ (\log \Lambda)^{-2}\left\{\frac{\gamma}{2}\log r + 2 \sum_{n=1}^{\infty} \frac{(\gamma + \log(p_n/2))J_0(p_n r)}{p_n^2 J_1(p_n)^2}\right\}$$

$$+ O((\log \Lambda)^{-3}). \tag{2.140}$$

As already noted Stewartson and Waechter (1976) showed that even the solutions due to Riley *et al.* (1974) have a singular behaviour near the centre as t^* tends to t_F^*. As the sphere or cylinder finally freezes, a boundary-layer structure develops, the thickness of which is of order $(\alpha\tau_e - t)^{1/2}$. At first this appears to be a remarkable phenomenon, because as time proceeds thermal boundary layers generally thicken rather than contract. The reason for the phenomenon can be traced to two facts. First, when a time $(\alpha\tau_e - t)$ is left to complete freezing, only those temperature variations that have a length scale of order $(\alpha\tau_e - t)^{1/2}$ can vary significantly in the time available. Secondly, significant spatial variations on this ever shortening length scale persist because the movement of the boundary toward the centre is linked with the thermal evolution in the solid, in the immediate neighbourhood of the interface, via the Stefan condition (2.6). For the sphere Stewartson and Waechter (1976) determined a new asymptotic solution valid for times

$$t^* = t_F^* + O(t_D^* e^{-(2\alpha)^{1/2}}). \tag{2.141}$$

For the cylinder Soward (1980) obtains the corresponding solution valid when

$$t^* = t_F^* + O(t_D^* \Lambda^{-\log \Lambda}). \tag{2.142}$$

These solutions only lead to small corrections to τ_e although the temperature profiles in the neighbourhood of the centre at the end time t_F^* are of some interest. We refer the reader to Soward (1980) for these details and for further discussion of the methods used by Riley *et al.* (1974) and by Stewartson and Waechter (1976).

For our purposes we may deduce from these developments the following estimates of the non-dimensional times to complete solidification for
the sphere and cylinder, respectively:

$$\hat{t}_c = \frac{\alpha + 1}{6} - \frac{2^{1/2}\zeta(3)}{\alpha^{1/2}\pi^{5/2}}, \tag{2.143}$$

$$\hat{t}_c = \frac{\alpha + 1}{4} - \frac{1}{4\log\Lambda} + \frac{1}{(\log\Lambda)^2}\left\{\frac{\gamma - 2\log 2}{8} + 2\sum_{n=1}^{\infty}\frac{\log p_n}{p_n^4 J_1(p_n)^2}\right\}, \tag{2.144}$$

where Λ is defined by (2.134) and in deriving (2.144) from (2.139) we
have utilized the asymptotic result

$$\lim_{r\to 0}\sum_{n=1}^{\infty}\frac{J_0(p_n r)}{p_n^4 J_1(p_n)^2} = \frac{1}{8}, \tag{2.145}$$

given in Appendix A of Soward (1980). Finally, we emphasize that the
boundary-layer analysis due to these authors applies only for large α. For
a small Stefan number the boundary-layer structure does not develop to
the same extent as is apparent from Figs 1.2 and 1.3.

2.7 Additional symbols used

$a_0(\bar{t}), a_1(\bar{t}), b_0(\bar{t}), b_1(\bar{t})$	coefficients in large α expansions of $A(t)$ and $B(t)$, (2.27)
$A(t), B(t)$	arbitrary functions of time, (2.11)
$c_n(z, z_0), e_n(z, z_0)$	functions in (2.23)
J_0, J_1	Bessel functions, (2.139)
$K(\alpha)$	Euler transformation parameter, (2.115)
p_n	nth positive zero of $J_0(p)$, (2.139)
r_0, x_0	arbitrary constants ($r_0 > 0$), (2.11)
$R_0(\bar{t}), R_1(\bar{t}), X_0(\bar{t}), X_1(\bar{t})$	coefficients in large α expansions of $R(t)$ and $X(t)$, (2.28)
S	Euler transformation variable, (2.115)
$\bar{t}$	stretched time t/α, (2.26)
$\bar{t}_i(X/\beta)$	coefficients in large α expansion of t, (2.78)
t_D^*	diffusion time, (2.129)
t_F^*	freezing or solidification time, (2.130)
u_0, u_1, u_2	coefficients in large α expansion of $u(\rho, \tau)$, (2.67)
v_1, v_3	coefficients in (2.114)
z, z_0	$r^2/4$ and $r_0^2/4$, respectively, (2.24)

Greek symbols

Λ	constant related to the Stefan number by (2.134)
τ_c, τ_s	constants involved in τ_e for cylinders and spheres, respectively, (2.138) and (2.135)
τ_e	constant involved in freezing time t_F^*, (2.130)
ϕ, Φ	stretched position and boundary position for sphere, (2.119)

Convention

Dot denotes differentiation with respect to stretched time $\bar{t} = t/\alpha$.

3

Integral formulation and bounds

3.1 Introduction

Many approximate solutions of Stefan problems are based on integral formulations. Some of these approximations are discussed in the following two chapters. In this chapter we present an integral formulation which is fundamental in the sense that it forms the basis of a number of other developments. Firstly, it leads to a formal integration of the motion of the boundary which, together with the inequalities $0 \leq T \leq 1$, yield simple upper and lower bounds for the motion. Secondly, from a Green's function derivation of this integral formulation it is apparent that the temperature T is bounded above by the pseudo-steady-state approximation T_{pss}, that is $T \leq T_{pss}$, and from this inequality an improved upper bound for the motion of the boundary is obtained. Thirdly, this integral formulation gives rise to formal series solutions and integral equations which are given in Chapter 4.

The basic integral formulations of the three freezing problems for slabs, spheres and cylinders may be formally derived in a variety of ways. In the following section we give the obvious derivation involving straightforward integration of the diffusion equation. We may also derive these results by multiplying the diffusion equation by an arbitrary function and then integrating. In Section 3.3 we derive the same results using the Green's function associated with the spatial component of the diffusion equation. Although for the three problems considered the three approaches are equivalent, this is not always the case and the Green's function approach is far more general. This aspect is discussed further in Chapter 7 for the problem involving both a fast and slow chemical reaction with governing equation (1.63). Moreover, the inequality $T \leq T_{pss}$ is immediately apparent from the Green's function derivation. The next three sections of the chapter deal with obtaining bounds for the motion of the moving boundary. In Section 3.4 we utilize $0 \leq T \leq 1$, while in Section 3.5 we use $T \leq T_{pss}$ to improve the upper bound. In Section 3.6 we show that the lower bound can also be improved by a further integration of the formal equation for the motion of the boundary. The bounds obtained are simple, therefore practically useful and although they apply for all values of the Stefan number α they are tightest for large α. Although it is not always easy to demonstrate a genuine bound, it is apparent that the general area is open to further improvements and improvisation of existing results. In particular it would be worth while establishing the bounding character, if any, of various standard approximations such as those examined in the following three chapters. Throughout the chapter we follow the practice of deriving the formulae for planar freezing and then simply stating the main results for spherical and cylindrical freezing. In the final section we summarize the results obtained in terms of λ in (1.34), which is 0, 1 or 2 for plane, cylindrical or spherical geometry, respectively.

The results given in this chapter are based on Dewynne and Hill (1984), Hill (1984a) and Hill and Dewynne (1984). However, similar ideas and techniques have been utilized by other authors but, because of different notations and terminologies, sometimes the correspondence is not altogether apparent. The reader may wish to consult the following more physically based papers which relate to this chapter: Glasser and Kern (1978), Kern (1977), Kern and Wells (1977), Hamill and Bankoff (1963) and El-Genk and Cronenberg (1979). For integral formulations which utilize the Green's function of the full diffusion equation we refer the reader to Chuang and Szekely (1971, 1972), Chuang and Ehrich (1974), Shaw (1974) and De Mey (1977). We note also that Wu (1966) obtains bounds for a multi-phase melting finite slab which is insulated on one face and is heated in an arbitrary manner on the other. The material is assumed to have a finite number of transformation temperatures and each phase may have distinct thermal properties.

3.2 Integral formulation by integration

In this section we give the basic integral formulations for the three solidification problems which we derive by two methods, both of which involve integrating the diffusion equation. For the problem of planar freezing defined by (2.1)–(2.3) we have, on integrating (2.1) from an arbitrary position x to the moving boundary $X(t)$,

$$\int_{x}^{X(t)} \frac{\partial T}{\partial t}(\xi, t) \, d\xi = -\alpha \frac{dX}{dt} - \frac{\partial T}{\partial x}(x, t), \qquad (3.1)$$

where we have used the Stefan condition $(2.3)_1$. On repeating this integration for (3.1) and using

$$\int_x^{X(t)} \int_\eta^{X(t)} \frac{\partial T}{\partial t} (\xi, t) \, d\xi \, d\eta = \int_x^{X(t)} \int_x^\xi \frac{\partial T}{\partial t} (\xi, t) \, d\eta \, d\xi$$

$$= \int_x^{X(t)} (\xi - x) \frac{\partial T}{\partial t} (\xi, t) \, d\xi, \qquad (3.2)$$

we obtain on noting $(2.2)_2$

$$\int_x^{X(t)} (\xi - x) \frac{\partial T}{\partial t} (\xi, t) \, d\xi = -\alpha \frac{dX}{dt} (X - x) + T(x, t), \qquad (3.3)$$

which can be rearranged to give the integro-partial differential equation

$$T(x, t) = \frac{\partial}{\partial t} \int_x^{X(t)} (\xi - x)[\alpha + T(\xi, t)] \, d\xi. \qquad (3.4)$$

We observe that in deriving (3.4) we have utilized all the Eqs. (2.1)–(2.3) except the surface condition $(2.2)_1$ and the initial condition $(2.3)_2$. From $(2.2)_1$ and (3.4) we have

$$\frac{d}{dt} \int_0^{X(t)} (\xi + \beta)[\alpha + T(\xi, t)] \, d\xi = 1, \qquad (3.5)$$

which evidently integrates immediately so that together with $(2.3)_2$ we conclude

$$t = \int_0^{X(t)} (\xi + \beta)[\alpha + T(\xi, t)] \, d\xi. \qquad (3.6)$$

Equations (3.4) and (3.6) constitute the basic integral formulation for planar freezing.

Although (3.6) is not an explicit equation for the motion of the boundary it nevertheless is an important result. First we observe that this equation is equivalent to

$$t = -\alpha \int_0^{X(t)} \left\{ \frac{\partial T^\dagger}{\partial x} (X, X) \right\}^{-1} dX, \qquad (3.7)$$

which follows from (2.3), where $T^\dagger(x, X)$ denotes the temperature $T(x, t)$ but with (x, X) as independent variables. The equality of (3.6) and (3.7) is by no means obvious. Equation (3.6) is an integral involving temperature values, while (3.7) involves flux values at the boundary. In order to see that these equations are indeed equivalent, we simply differentiate (3.6) with respect to time so that we have

$$1 = \alpha(X + \beta) \frac{dX}{dt} + \int_0^{X(t)} (\xi + \beta) \frac{\partial^2 T}{\partial \xi^2} (\xi, t) \, d\xi, \qquad (3.8)$$

where we have used the diffusion equation (2.1). On integration by parts, (3.8) becomes

$$1 = (X + \beta)\left\{\alpha \frac{\mathrm{d}X}{\mathrm{d}t} + \frac{\partial T}{\partial x}(X(t), t)\right\} + T(0, t) - \beta \frac{\partial T}{\partial x}(0, t), \qquad (3.9)$$

which on using $(2.2)_1$ yields precisely $(2.3)_1$, which is equivalent to (3.7). Clearly, (3.6) embodies more information than does (3.7) and may be used to advantage in conjunction with other approximate procedures. For example, using (3.6) for the large Stefan number expansions given in the previous chapter, we may deduce either (2.75) or (2.77) from simply u_0 and u_1, while using (3.7) requires a knowledge of u_0, u_1 and u_2 and involves a more difficult calculation.

In order to show that (3.4) and (3.6) are consistent with the exact Neumann solution we have from (1.49) for the integral in (3.4)

$$\alpha \int_x^{X(t)} (\xi - x)\left\{1 + \gamma \int_{\xi/X(t)}^1 e^{\gamma(1-\eta^2)/2}\,\mathrm{d}\eta\right\}\mathrm{d}\xi$$

$$= \alpha X^2 \int_\rho^1 (\omega - \rho)\left\{1 + \gamma \int_\omega^1 e^{\gamma(1-\eta^2)/2}\,\mathrm{d}\eta\right\}\mathrm{d}\omega$$

$$= \frac{\alpha X^2}{2}\left\{(1 - \rho)^2 + \gamma \int_\rho^1 (\eta - \rho)^2 e^{\gamma(1-\eta^2)/2}\,\mathrm{d}\eta\right\}$$

$$= \frac{\alpha X^2}{2}\left\{\rho^2 - \rho\, e^{\gamma(1-\rho^2)/2} + (1 + \gamma\rho^2)\int_\rho^1 e^{\gamma(1-\eta^2)/2}\,\mathrm{d}\eta\right\}, \qquad (3.10)$$

where $\rho = x/X(t)$, $\omega = \xi/X(t)$ and we have changed orders of integration and used integration by parts in a routine manner to deduce (3.10). On differentiating the expression (3.10) partially with respect to time, Eq. (3.4) becomes

$$T(x, t) = \alpha X \frac{\mathrm{d}X}{\mathrm{d}t} \int_\rho^1 e^{\gamma(1-\eta^2)/2}\,\mathrm{d}\eta. \qquad (3.11)$$

In a similar manner, (3.6) with β zero becomes

$$t = \frac{\alpha X^2}{2} \int_0^1 e^{\gamma(1-\eta^2)/2}\,\mathrm{d}\eta, \qquad (3.12)$$

and these equations are clearly consistent with (1.42) and (1.43).

We may give a slightly different derivation of (3.4) by multiplying the diffusion equation (2.1) by an arbitrary function $\phi(x, \xi, t)$, such that $\phi(x, X, t)$ is non-zero and integrating once from x to $X(t)$. For the left-hand side we have

$$\int_x^{X(t)} \phi(x, \xi, t) \frac{\partial T}{\partial t}(\xi, t)\, d\xi = \frac{\partial}{\partial t} \int_x^{X(t)} \phi(x, \xi, t) T(\xi, t)\, d\xi$$

$$- \int_x^{X(t)} T(\xi, t) \frac{\partial \phi}{\partial t}(x, \xi, t)\, d\xi, \qquad (3.13)$$

while for the right-hand side we obtain

$$\int_x^{X(t)} \phi(x, \xi, t) \frac{\partial^2 T}{\partial \xi^2}(\xi, t)\, d\xi = -\alpha \frac{dX}{dt} \phi(x, X, t)$$

$$- \frac{\partial T}{\partial x}(x, t)\phi(x, x, t)$$

$$+ T(x, t) \frac{\partial \phi}{\partial \xi}(x, x, t)$$

$$+ \int_x^{X(t)} T(\xi, t) \frac{\partial^2 \phi}{\partial \xi^2}(x, \xi, t)\, d\xi. \qquad (3.14)$$

From these equations we readily deduce

$$T(x, t) \frac{\partial \phi}{\partial \xi}(x, x, t) - \frac{\partial T}{\partial x} \phi(x, x, t)$$

$$= \frac{\partial}{\partial t} \int_x^{X(t)} \phi(x, \xi, t)[\alpha + T(\xi, t)]\, d\xi$$

$$- \int_x^{X(t)} T(\xi, t) \left\{ \frac{\partial \phi}{\partial t}(x, \xi, t) + \frac{\partial^2 \phi}{\partial \xi^2}(x, \xi, t) \right\} d\xi, \qquad (3.15)$$

and we observe that (3.4) follows immediately with

$$\phi(x, \xi, t) = \xi - x. \qquad (3.16)$$

It is worth while noting that we may establish the inequality $T \leq T_{pss}$ for planar freezing by eliminating t from (3.4) and (3.5) so that

$$T^\dagger(x, X) = \frac{\dfrac{\partial}{\partial X}\left(\displaystyle\int_x^X (\xi - x)[\alpha + T^\dagger(\xi, X)]\, d\xi \right)}{\dfrac{\partial}{\partial X}\left(\displaystyle\int_0^X (\xi + \beta)[\alpha + T^\dagger(\xi, X)]\, d\xi \right)}, \qquad (3.17)$$

and thus

$$T^\dagger(x, X) - \frac{X-x}{X+\beta}$$

$$= -\frac{\left\{ (X-x) \int_0^x (\xi + \beta) \frac{\partial T^\dagger}{\partial X}(\xi, X)\, d\xi \atop + (x + \beta) \int_x^X (X - \xi) \frac{\partial T^\dagger}{\partial X}(\xi, X)\, d\xi \right\}}{\alpha(X+\beta) + \int_0^X (\xi + \beta) \frac{\partial T^\dagger}{\partial X}(\xi, X)\, d\xi}. \tag{3.18}$$

Now both $\partial T/\partial t \geq 0$ and $dX/dt \geq 0$ and therefore

$$\frac{\partial T^\dagger}{\partial X} = \frac{\partial T}{\partial t} \Big/ \frac{dX}{dt} \geq 0, \tag{3.19}$$

and the inequality $T \leq T_{\text{pss}}$ follows. This result is established more directly in the following section using a Green's function formulation. We remark that in Chapter 4 we utilize (3.17) in an integral iteration procedure. Finally in this section we give the integral formulations for spherical and cylindrical freezing.

3.2.1 Spherical freezing.
On multiplying (2.4) by r^2 and utilizing $(2.5)_2$ and $(2.6)_1$ we may deduce, in an entirely analogous manner, the following:

$$\frac{\partial T}{\partial r}(r, t) = \frac{\partial}{\partial t} \int_{R(t)}^r \left(\frac{\xi}{r}\right)^2 [\alpha + T(\xi, t)]\, d\xi, \tag{3.20}$$

$$T(r, t) = \frac{\partial}{\partial t} \int_{R(t)}^r \xi^2 \left(\frac{1}{\xi} - \frac{1}{r}\right)[\alpha + T(\xi, t)]\, d\xi, \tag{3.21}$$

and from these equations and $(2.5)_1$ we have

$$1 = \frac{d}{dt} \int_{R(t)}^1 \xi[1 + (\beta - 1)\xi][\alpha + T(\xi, t)]\, d\xi. \tag{3.22}$$

On integration and noting $(2.6)_2$ we obtain

$$t = \int_{R(t)}^1 \xi[1 + (\beta - 1)\xi][\alpha + T(\xi, t)]\, d\xi, \tag{3.23}$$

and (3.21) and (3.23) constitute the basic equations for spherical freezing.

3.2.2 Cylindrical freezing.
Similarly, on multiplying (2.7) by r and

integrating from $R(t)$ to r we obtain

$$\frac{\partial T}{\partial r}(r, t) = \frac{\partial}{\partial t} \int_{R(t)}^{r} \frac{\xi}{r} [\alpha + T(\xi, t)] \, d\xi, \tag{3.24}$$

$$T(r, t) = \frac{\partial}{\partial t} \int_{R(t)}^{r} \xi(\log r - \log \xi)[\alpha + T(\xi, t)] \, d\xi. \tag{3.25}$$

From the surface boundary condition and these equations we deduce

$$1 = \frac{d}{dt} \int_{R(t)}^{1} \xi(\beta - \log \xi)[\alpha + T(\xi, t)] \, d\xi, \tag{3.26}$$

and therefore as before the boundary motion is determined from

$$t = \int_{R(t)}^{1} \xi(\beta - \log \xi)[\alpha + T(\xi, t)] \, d\xi. \tag{3.27}$$

We note that we have purposely noted the intermediate equations (3.22) and (3.26), since the important inequalities (3.81) and (3.89) follow immediately from these equations using $\partial T/\partial t \geq 0$.

3.3 Integral formulation by Green's function

We first note some mathematical preliminaries for Green's functions. Consider the self-adjoint problem with homogeneous boundary conditions,

$$\frac{d}{dx}\left(p(x)\frac{dy}{dx}\right) + q(x)y = f(x), \qquad a < x < b, \tag{3.28}$$

$$y(a) + \lambda_1 \frac{dy}{dx}(a) = 0, \qquad y(b) + \lambda_2 \frac{dy}{dx}(b) = 0, \tag{3.29}$$

where $p(x)$, $q(x)$ and $f(x)$ denote arbitrary functions of x such that $p(x)$ is non-zero in the interval $[a, b]$ and λ_1 and λ_2 denote arbitrary real constants. Assuming there is no solution of the homogeneous equation

$$\frac{d}{dx}\left(p(x)\frac{dy}{dx}\right) + q(x)y = 0, \tag{3.30}$$

which satisfies both homogeneous boundary conditions, we let $y_1(x)$ and $y_2(x)$ denote any linearly independent solutions of (3.30) such that $y_1(x)$ satisfies the first boundary condition of (3.29) and $y_2(x)$ satisfies the second. We may readily verify that the solution of (3.28) and (3.29) is given by

$$y(x) = \int_a^b G(x, \xi) f(\xi) \, d\xi, \tag{3.31}$$

where the symmetric Green's function $G(x, \xi)$ is defined by

$$G(x, \xi) = \frac{y_1(\xi) y_2(x)}{w_0}, \qquad a \leq \xi \leq x,$$

$$= \frac{y_1(x) y_2(\xi)}{w_0}, \qquad x \leq \xi \leq b, \tag{3.32}$$

where w_0 is a constant determined from the equation

$$p(x)\left(y_1(x) \frac{dy_2(x)}{dx} - y_2(x) \frac{dy_1(x)}{dx}\right) = w_0. \tag{3.33}$$

Using the Green's function approach we establish the same results given in the previous section but in the reverse order. That is, $T \leq T_{pss}$ emerges first, then the formal integral (3.6) for the motion of the boundary and finally the integral formulation (3.4) is proved. For planar freezing we let

$$u(x, t) = T(x, t) - T_{pss}(x, t), \tag{3.34}$$

where the pseudo-steady-state approximation is defined by (1.66) so that from (2.1) and (2.2) we obtain

$$\frac{\partial^2 u}{\partial x^2} = \frac{\partial T}{\partial t}, \qquad 0 < x < X(t), \tag{3.35}$$

$$u(0, t) - \beta \frac{\partial u}{\partial x}(0, t) = 0, \qquad u(X(t), t) = 0. \tag{3.36}$$

For this problem with

$$u_1(x, t) = x + \beta, \qquad u_2(x, t) = x - X, \tag{3.37}$$

we have from (3.31)–(3.33)

$$T(x, t) = T_{pss}(x, t) + \int_0^X G(x, \xi, t) \frac{\partial T}{\partial t}(\xi, t) \, d\xi, \tag{3.38}$$

where the Green's function $G(x, \xi, t)$ is given by

$$G(x, \xi, t) = \frac{(\xi + \beta)(x - X)}{(X + \beta)}, \qquad 0 \leq \xi \leq x,$$

$$= \frac{(x + \beta)(\xi - X)}{(X + \beta)}, \qquad x \leq \xi \leq X. \tag{3.39}$$

Now since $G \leq 0$ and $\partial T / \partial t \geq 0$ it follows immediately from (3.38) that $T \leq T_{pss}$. Moreover, from (1.66) and the above equations we have

$$\frac{\partial T}{\partial x}(x, t) = -\frac{1}{X + \beta} + \int_0^x \frac{\xi + \beta}{X + \beta} \frac{\partial T}{\partial t}(\xi, t)\, d\xi + \int_x^X \frac{\xi - X}{X + \beta} \frac{\partial T}{\partial t}(\xi, t)\, d\xi,$$

$$(3.40)$$

and from this equation and (2.3)$_1$ we deduce

$$-\frac{1}{X + \beta} + \int_0^X \frac{\xi + \beta}{X + \beta} \frac{\partial T}{\partial t}(\xi, t)\, d\xi = -\alpha \frac{dX}{dt}, \tag{3.41}$$

which may be rearranged to give (3.5) and hence the formal integral for the motion (3.6). In order to deduce (3.4) from (3.38) we use (3.5) to write (3.38) as follows:

$$T(x, t) = \frac{X - x}{X + \beta} \frac{d}{dt} \int_0^X (\xi + \beta)[\alpha + T(\xi, t)]\, d\xi$$

$$+ \int_0^x \frac{(\xi + \beta)(x - X)}{X + \beta} \frac{\partial T}{\partial t}(\xi, t)\, d\xi$$

$$+ \int_x^X \frac{(x + \beta)(\xi - X)}{X + \beta} \frac{\partial T}{\partial t}(\xi, t)\, d\xi, \tag{3.42}$$

which simplifies to give

$$T(x, t) = (X - x)\alpha \frac{dX}{dt} + \int_x^X (\xi - x) \frac{\partial T}{\partial t}(\xi, t)\, d\xi, \tag{3.43}$$

and (3.4) is therefore established. Although we have considered a particularly simple example, the above procedure and structure are quite general provided we ensure that the spatial component of the diffusion equation is in self-adjoint form. We now note the corresponding results for spherical and cylindrical freezing.

3.3.1 Spherical freezing.

Again we introduce $u(r, t)$ such that

$$u(r, t) = T(r, t) - T_{pss}(r, t), \tag{3.44}$$

where the pseudo-steady-state estimate is given by (1.70) and from (2.4) and (2.5) we have

$$\frac{\partial}{\partial r}\left(r^2 \frac{\partial u}{\partial r}\right) = r^2 \frac{\partial T}{\partial t}, \qquad R(t) < r < 1, \tag{3.45}$$

$$u(R(t), t) = 0, \qquad u(1, t) + \beta \frac{\partial u}{\partial r}(1, t) = 0. \tag{3.46}$$

For the self-adjoint spatial operator in (3.45) the functions corresponding to $y_1(x)$ and $y_2(x)$ are given by

$$u_1(r, t) = \frac{1}{R} - \frac{1}{r}, \qquad u_2(r, t) = \frac{1}{r} - 1 + \beta, \tag{3.47}$$

so that the Green's function becomes

$$G(r, \xi, t) = -\frac{(1/R - 1/\xi)(1/r - 1 + \beta)}{(1/R - 1 + \beta)}, \qquad R \leq \xi \leq r,$$

$$= -\frac{(1/R - 1/r)(1/\xi - 1 + \beta)}{(1/R - 1 + \beta)}, \qquad r \leq \xi \leq 1, \tag{3.48}$$

and we have

$$T(r, t) = T_{pss}(r, t) + \int_R^1 G(r, \xi, t)\xi^2 \frac{\partial T}{\partial t}(\xi, t)\,d\xi. \tag{3.49}$$

Again assuming $\beta \geq 0$ we may show $T \leq T_{pss}$ and (3.23) and (3.21) follow exactly as described for planar freezing.

3.3.2 Cylindrical freezing. In this case with $u(r, t)$ defined by (3.44) we have in place of (3.45)

$$\frac{\partial}{\partial r}\left(r\frac{\partial u}{\partial r}\right) = r\frac{\partial T}{\partial t}, \qquad R(t) < r < 1, \tag{3.50}$$

while (3.46) is unaltered. The appropriate linearly independent solutions of the homogeneous problem are given by

$$u_1(r, t) = \log r - \log R, \qquad u_2(r, t) = \log r - \beta, \tag{3.51}$$

so that the Green's function becomes

$$G(r, \xi, t) = -\frac{(\log \xi - \log R)(\beta - \log r)}{(\beta - \log R)}, \qquad R \leq \xi \leq r,$$

$$= -\frac{(\log r - \log R)(\beta - \log \xi)}{(\beta - \log R)}, \qquad r \leq \xi \leq 1. \tag{3.52}$$

The relation (3.49) holds with T_{pss} given by (1.73) and ξ in place of ξ^2 in the integral.

Finally in this section we make one or two comments relating to the above. Firstly, for spherical and cylindrical freezing problems but with boundaries moving outward in an infinite region, the pseudo-steady-state approximation T_{pss} is still an upper bound. In this situation the problems are still defined by (2.4)–(2.6) and (2.7) but, with $\beta \leq 0$ and the Green's functions as given by (3.48) and (3.52), although positive are integrated over a negative interval in (3.49) and therefore $T \leq T_{pss}$. (This corrects a statement in Hill (1984a).) Secondly, in the above we have frequently utilized the condition that the temperature is zero on the moving front. For example, for spherical freezing if $(2.5)_2$ is replaced by (1.62) so that T_{pss} is given by (1.75) then (3.49) still holds where the Green's function is given by

$$G(r, \xi, t) = -\frac{(1/R - 1/\xi + \delta/R^2)(1/r - 1 + \beta)}{(1/R - 1 + \beta + \delta/R^2)}, \qquad R \le \xi \le r,$$

$$= -\frac{(1/R - 1/r + \delta/R^2)(1/\xi - 1 + \beta)}{(1/R - 1 + \beta + \delta/R^2)}, \qquad r \le \xi \le 1. \qquad (3.53)$$

Moreover, in place of (3.21) and (3.23) we have, respectively,

$$T(r, t) = \frac{\partial}{\partial t} \int_{R(t)}^r \xi^2 \left(\frac{1}{\xi} - \frac{1}{r} + \frac{\delta}{\xi^2}\right) [\alpha + T(\xi, t) - T(\xi, t_\xi)] \, d\xi, \qquad (3.54)$$

$$t = \alpha \delta [1 - R(t)] + \int_{R(t)}^1 \xi [1 + (\beta - 1)\xi][\alpha + T(\xi, t) - T(\xi, t_\xi)] \, d\xi, \qquad (3.55)$$

where t_ξ is defined by $\xi = R(t_\xi)$. Thus the effect of T non-zero on the moving boundary is to introduce additional terms in the basic integral formulation. Thirdly, we note that standard transient heat-conduction problems also admit similar integral formulations to the above, although they are seldom, if at all, utilized. For example, for the hollow sphere $b < r < 1$ with zero initial temperature and surfaces $r = 1$ and $r = b$ held at constant temperatures unity and zero, respectively, we have in place of (3.21) and (3.23)

$$T(r, t) = \frac{\partial}{\partial t} \int_b^r \xi^2 \left(\frac{1}{\xi} - \frac{1}{r}\right) T(\xi, t) \, d\xi + b^2 \left(\frac{1}{b} - \frac{1}{r}\right) \frac{\partial T}{\partial r}(b, t), \qquad (3.56)$$

$$t = \int_b^1 \xi(1 - \xi) T(\xi, t) \, d\xi + b(1 - b) \int_0^t \frac{\partial T}{\partial r}(b, v) \, dv. \qquad (3.57)$$

We may show that the exact solution (see Carslaw and Jaeger 1965, p. 246)

$$T(r, t) = \frac{(b - r)}{r(b - 1)} + \frac{2}{\pi r} \sum_{n=1}^\infty \frac{1}{n} \exp\left(-\frac{n^2 \pi t}{(b - 1)^2}\right) \sin \frac{n\pi(r - 1)}{(b - 1)}, \qquad (3.58)$$

satisfies these relations identically.

3.4 Bounds on the motion from $0 \le T \le 1$

To demonstrate the usefulness of the formal integrals (3.6), (3.23) and (3.27) for the motion of the plane, spherical and cylindrical boundaries, respectively, we show in this section that even the crude inequalities $0 \le T \le 1$ give rise to non-trivial bounds on the motion. First, for planar freezing we have from (3.6) and $0 \le T \le 1$ the inequalities

$$\frac{\alpha}{2} X(X + 2\beta) \le t \le \frac{\alpha + 1}{2} X(X + 2\beta). \tag{3.59}$$

The lower bound is merely the pseudo-steady-state estimate of the motion, while the upper bound is clearly a constant multiple of the pseudo-steady-state motion. Thus the pseudo-steady-state boundary moves faster than the actual boundary. For β zero we have from the exact motion $X(t) = (2\gamma t)^{1/2}$ and (3.59)

$$\alpha \le \gamma^{-1} \le \alpha + 1, \tag{3.60}$$

so that the upper bound is not as tight as (1.59). In the following section we show that precisely the upper bound in (1.59) arises from $T \le T_{pss}$.

Similarly, for spherical freezing we have from (3.23) and $0 \le T \le 1$

$$\frac{\alpha}{6} (1 - R)[(1 + 2\beta)(1 + R) + 2(\beta - 1)R^2] \le t$$

$$\le \frac{\alpha + 1}{6} (1 - R)[(1 + 2\beta)(1 + R) + 2(\beta - 1)R^2], \quad (3.61)$$

so that in particular the time t_c to complete freezing satisfies the inequalities

$$\frac{\alpha}{6} (1 + 2\beta) \le t_c \le \frac{\alpha + 1}{6} (1 + 2\beta). \tag{3.62}$$

These simple results may be useful either from a practical viewpoint or in conjunction with estimating appropriate step sizes in a numerical scheme. For cylindrical freezing we obtain from (3.27) and $0 \le T \le 1$

$$\frac{\alpha}{4} [2R^2 \log R + (1 + 2\beta)(1 - R^2)] \le t$$

$$\le \frac{\alpha + 1}{4} [2R^2 \log R + (1 + 2\beta)(1 - R^2)], \quad (3.63)$$

and therefore t_c satisfies

$$\frac{\alpha}{4} (1 + 2\beta) \le t_c \le \frac{\alpha + 1}{4} (1 + 2\beta). \tag{3.64}$$

Although for the motion of the boundaries we subsequently improve both the upper and lower bounds, we have purposely noted the above estimates to emphasize the utility of the formal integrals (3.6), (3.23) and

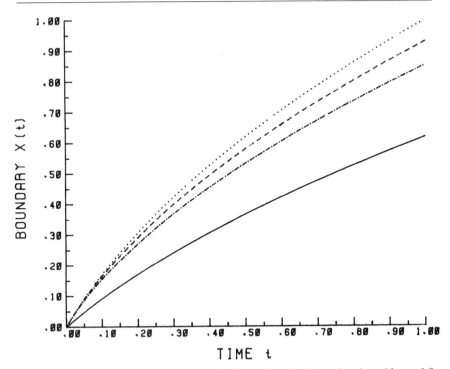

Figure 3.1 Various upper and lower bounds for motion for planar freezing with $\alpha = 1.0$ and $\beta = 0.5$ ($\cdots\cdots$ pseudo-steady-state (3.59), $----$ improved lower bound (3.76), $-\cdot-\cdot-\cdot-$ improved upper bound (3.66), ——— upper bound (3.59))

(3.27). Indeed, roughly speaking, any reasonable estimate of the temperature T together with these integrals gives rise to meaningful expressions for the motion. The various upper and lower bounds for plane, spherical and cylindrical freezing are shown graphically in Figs 3.1–3.3, respectively. Numerical values of the various estimates of γ^{-1} for the exact Neumann solution for β zero are given in Table 3.1. Numerical values of the various bounds for t_c are given in Tables 3.2 and 3.3 for spherical and cylindrical freezing, respectively.

3.5 Improved upper bound from $T \le T_{pss}$

In this section we give the improved upper bounds to the motion which result from the formal integrals (3.6), (3.23) and (3.27) and $T \le T_{pss}$. For planar, spherical and cylindrical solidification these upper bounds turn out to be precisely the order one corrected motions for large α given previously by Eqs (2.41), (2.51) and (2.61), respectively. Although for

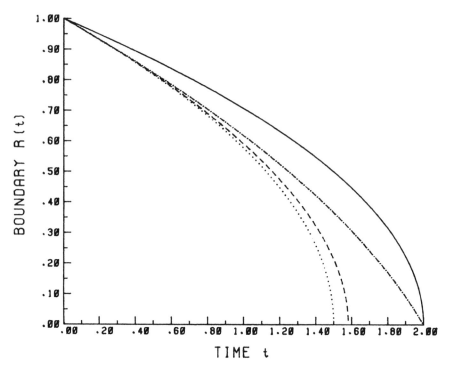

Figure 3.2 Various upper and lower bounds for motion for spherical freezing with $\alpha = 3.0$ and $\beta = 1.0$ ($\cdots$ pseudo-steady-state (3.61), $----$ improved lower bound (3.84), $-\cdot-\cdot-$ improved upper bound (3.70), ——— upper bound (3.61))

spherical and cylindrical solidification the bounds (3.70) and (3.72) represent genuine improvements on the upper bounds in (3.61) and (3.63), unfortunately, however, these bounds coincide with those given previously for R zero. Thus the inequalities (3.62) and (3.64) are unchanged.

For planar freezing we have from (3.6) and (1.66)

$$t \leq \int_0^X (\xi + \beta)\left\{\alpha + \left(\frac{X - \xi}{X + \beta}\right)\right\} d\xi, \tag{3.65}$$

which on evaluating the integral becomes

$$t \leq \frac{\alpha X}{2}(X + 2\beta) + \frac{X^2(X + 3\beta)}{6(X + \beta)}. \tag{3.66}$$

Thus the first two terms of either (2.41) or (2.77) constitute an upper bound to the motion of the boundary. For the exact Neumann solution with β zero we have from $X(t) = (2\gamma t)^{1/2}$ and (3.66)

$$\gamma^{-1} \leq \alpha + \tfrac{1}{3}, \tag{3.67}$$

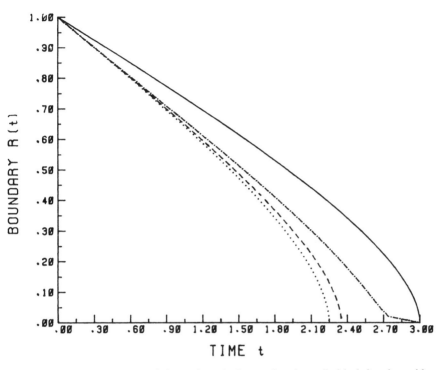

Figure 3.3 Various upper and lower bounds for motion for cylindrical freezing with $\alpha = 3.0$ and $\beta = 1.0$ ($\cdots\cdots$ pseudo-steady-state (3.63), $-----$ improved lower bound (3.91), $-\cdot-\cdot-\cdot-$ improved upper bound (3.72), $\underline{\qquad}$ upper bound (3.63))

Table 3.1 Numerical values of various bounds for γ^{-1} for planar freezing for $\beta = 0$ (from Hill and Dewynne (1984))

	Lower bounds		Exact solution γ^{-1}	Upper bound $\alpha + \frac{1}{3}$
α	$(3.78)_1$	$(3.78)_2$		
0.20	0.33	0.33	0.45	0.53
0.25	0.38	0.38	0.50	0.58
0.50	0.65	0.62	0.78	0.83
1.00	1.15	1.12	1.30	1.33
2.00	2.16	2.12	2.31	2.33
5.00	5.16	5.11	5.33	5.33
10.00	10.17	10.11	10.33	10.33
50.00	50.17	50.11	50.33	50.33
100.00	100.17	100.11	100.33	100.33
500.00	500.17	500.11	500.33	500.33

Table 3.2 Numerical values of various bounds for t_c for spherical freezing for $\beta = 1$ (from Hill and Dewynne (1984))

	Lower bounds			Upper bound
α	(3.62)	(3.87)	(3.88)	(3.62)
0.01	0.01	0.03	0.04	0.51
0.10	0.05	0.10	0.11	0.55
0.50	0.25	0.32	0.30	0.75
1.00	0.50	0.58	0.54	1.00
2.00	1.00	1.08	1.04	1.50
5.00	2.50	2.58	2.54	3.00
10.00	5.00	5.08	5.03	5.50
50.00	25.00	25.08	25.03	25.50
100.00	50.00	50.08	50.03	50.50
500.00	250.00	250.08	250.03	250.50

Table 3.3 Numerical values of various bounds for t_c for cylindrical freezing for $\beta = 1$ (from Hill and Dewynne (1984))

	Lower bounds			Upper bound
α	(3.64)	(3.93)	(3.94)	(3.64)
0.01	0.01	0.04	0.06	0.76
0.10	0.08	0.15	0.14	0.83
0.50	0.38	0.47	0.43	1.13
1.00	0.75	0.85	0.80	1.50
2.00	1.50	1.60	1.55	2.25
5.00	3.75	3.85	3.80	4.50
10.00	7.50	7.60	7.55	8.25
50.00	37.50	37.60	37.55	38.25
100.00	75.00	75.10	75.05	75.75
500.00	375.00	375.10	375.05	375.75

so that altogether the inequalities (1.59) are equivalent to

$$0 \le T \le T_{pss}. \tag{3.68}$$

Similarly, for spherical freezing we have from (3.23) and (1.70)

$$t \le \int_R^1 \xi[1 + (\beta - 1)\xi]\left\{\alpha + \frac{1 - R/\xi}{1 + (\beta - 1)R}\right\} d\xi, \tag{3.69}$$

which simplifies to give

$$t \le \frac{\alpha}{6}(1 - R)[(1 + 2\beta)(1 + R) + 2(\beta - 1)R^2]$$
$$+ \frac{(1 - R)^2[1 + 2\beta + (\beta - 1)R]}{6[1 + (\beta - 1)R]}, \tag{3.70}$$

and this coincides with the first two terms given in either (2.51) or (2.93). For cylindrical solidification we obtain from (3.27) and (1.73)

$$t \le \int_R^1 \xi(\beta - \log \xi)\left\{\alpha + \left(\frac{\log \xi - \log R}{\beta - \log R}\right)\right\} d\xi, \tag{3.71}$$

which gives

$$t \le \frac{\alpha}{4}[2R^2 \log R + (1 + 2\beta)(1 - R^2)]$$
$$+ \frac{1}{4}\left\{(1 + 2\beta + R^2) - \frac{1 + 2\beta + 2\beta^2 - R^2}{(\beta - \log R)}\right\}, \tag{3.72}$$

and again this coincides with the first two terms of either (2.61) or (2.107). We observe that (3.70) and (3.72) give rise to the same upper bounds in (3.62) and (3.64), respectively. For cylindrical freezing we note the perculiar behaviour of (3.72) shown in Fig. 3.3 as R tends to zero.

3.6   Improved lower bounds by integration

For plane freezing we may improve the pseudo-steady-state lower bound for the motion by further integrations of (3.6) and utilizing the inequality

$$\frac{1}{\dot{X}(t)} \ge \alpha(X + \beta), \tag{3.73}$$

where the dot denotes differentiation with respect to time and (3.73) follows from (3.5) noting that $\partial T/\partial t \ge 0$. From (3.6) on using (3.4) we have

$$t = t_{pss}(X) + \frac{d}{dt}\int_0^X \int_\xi^X (\xi + \beta)(\eta - \xi)[\alpha + T(\eta, t)] \, d\eta \, d\xi, \tag{3.74}$$

where $t_{pss}(X)$ is defined by (1.67) and which, on integrating with respect to time, gives

$$\frac{t^2}{2} = \int_0^t t_{pss}(X(v)) \, dv + \int_0^{X(t)} \int_\xi^{X(t)} (\xi + \beta)(\eta - \xi)[\alpha + T(\eta, t)] \, d\eta \, d\xi. \tag{3.75}$$

Now on using (3.73) for the first integral and $T \geq 0$ in the second we may deduce

$$t^2 \geq \frac{\alpha^2 X^2}{4}(X + 2\beta)^2 + \frac{\alpha}{12} X^3(X + 4\beta), \tag{3.76}$$

which is clearly an improvement on the pseudo-steady-state lower bound. In principle, we may continue this process and use (3.4) to effect a further integration of (3.75). However, it is apparent from numerical results that for all but the smallest values of α and β we obtain inferior results to (3.76). We might expect this phenomenon since the amount 'given away' by inequality (3.73) would eventually exceed the improvements obtained by substituting (3.4) at each step. For completeness a further integration of (3.75) using (3.4) and the inequalities (3.73) and $T \geq 0$ yields

$$t^3 \geq \frac{\alpha^3}{8} X^3 (X + 2\beta)^3 + \frac{\alpha^2}{24} X^4(X^2 + 6\beta X + 6\beta^2) + \frac{\alpha}{120} X^5(X + 6\beta). \tag{3.77}$$

For the exact Neumann solution $X(t) = (2\gamma t)^{1/2}$ and (3.76) and (3.77) with β zero we obtain, respectively, the inequalities

$$\frac{1}{\gamma} \geq \left(\alpha^2 + \frac{\alpha}{3}\right)^{1/2}, \qquad \frac{1}{\gamma} \geq \left(\alpha^3 + \frac{\alpha^2}{3} + \frac{\alpha}{15}\right)^{1/3}, \tag{3.78}$$

which are verified numerically in Table 3.1. It is apparent from this table that the new lower bounds are both significant improvements on the pseudo-steady-state estimate. It is also apparent that the second improvement (3.77) for the given values of α is inferior to the first improvement given by (3.76). It is perhaps worth noting that the constants appearing in the sequence of bounds (3.78) are the same constants arising in the expansion (1.57) of α^{-1} in powers of γ. In fact, we may show by induction for β zero that for any $n \geq 1$ we have

$$t^n \geq \frac{X^{2n}}{2^n}\left(\alpha^n + \frac{\alpha^{n-1}}{3} + \ldots + \frac{2^{n-1}(n-1)! \, \alpha}{(2n-1)!}\right), \tag{3.79}$$

so that in particular for any $n \geq 1$ we have

$$\frac{1}{\gamma} \geq \left(\alpha^n + \frac{\alpha^{n-1}}{3} + \ldots + \frac{2^{n-1}(n-1)! \, \alpha}{(2n-1)!}\right)^{1/n}. \tag{3.80}$$

However, numerical results indicate that successive bounds generally decrease with increasing n. That is, for all but the smallest values of α, $n = 2$ gives the tightest lower bound. Thus, even if corresponding general bounds for β non-zero could be established there is no reason to believe these would substantially improve (3.76). Figure 3.1 illustrates the extreme bounds (3.59) together with the improved upper and lower bounds

(3.66) and (3.76), respectively. The improvement (3.77) due to two integrations of (3.6) is not shown.

3.6.1 Spherical freezing.

In this case to improve the pseudo-steady-state lower bound we utilize the inequality

$$-\frac{1}{\dot{R}(t)} \geq \alpha R[1 + (\beta - 1)R], \tag{3.81}$$

which follows from (3.22) and $\partial T/\partial t \geq 0$. From (3.21) and (3.23) we have

$$t = t_{\text{pss}}(R) + \frac{d}{dt} \int_R^1 \int_R^\xi \eta[1 + (\beta - 1)\xi](\xi - \eta)[\alpha + T(\eta, t)] \, d\eta \, d\xi, \tag{3.82}$$

where $t_{\text{pss}}(R)$ is defined by (1.71). On integrating this equation and utilizing the inequalities (3.81) and $T \geq 0$ we have

$$t^2 \geq t_{\text{pss}}^2(R) + 2\alpha \int_R^1 \int_R^\xi \eta[1 + (\beta - 1)\xi](\xi - \eta) \, d\eta \, d\xi, \tag{3.83}$$

which becomes

$$t^2 \geq \frac{\alpha^2}{36}(1 - R)^2[(1 + 2\beta)(1 + R) + 2(\beta - 1)R^2]^2$$

$$+ \frac{\alpha}{60}(1 - R)^3[(1 - R)(1 + 4R) + 4\beta(1 + 3R + R^2)]. \tag{3.84}$$

A further integration of (3.82) and using the same inequalities gives

$$t^3 \geq t_{\text{pss}}^3(R) + 6\alpha^2 \int_R^1 \int_\zeta^1 \int_\zeta^\xi \eta\zeta[1 + (\beta - 1)\xi](\xi - \eta)[1 + (\beta - 1)\zeta] \, d\zeta \, d\eta \, d\xi$$

$$+ 6\alpha \int_R^1 \int_R^\xi \int_R^\eta \zeta[1 + (\beta - 1)\xi](\xi - \eta)(\eta - \zeta) \, d\zeta \, d\eta \, d\xi, \tag{3.85}$$

which yields

$$t^3 \geq \frac{\alpha^3}{216}(1 - R)^3[(1 + 2\beta)(1 + R) + 2(\beta - 1)R^2]^3$$

$$+ \frac{\alpha^2}{840}(1 - R)^4\Big\{(1 - R)^2(21R^2 + 12R + 2)$$

$$+ 6\beta(1 - R)(7R^3 + 16R^2 + 10R + 2)$$

$$+ 7\beta^2(3R^4 + 12R^3 + 10R^2 + 4R + 1)\Big\}$$

$$+ \frac{\alpha}{840}(1 - R)^5[(1 - R)(1 + 6R) + 6\beta(1 + 5R + R^2)]. \tag{3.86}$$

In particular, for the time t_c to complete solidification we have from (3.84) and (3.86) the following inequalities

$$t_c^2 \geq \frac{\alpha^2}{36}(1 + 2\beta)^2 + \frac{\alpha}{60}(1 + 4\beta), \tag{3.87}$$

$$t_c^3 \geq \frac{\alpha^3}{216}(1 + 2\beta)^3 + \frac{\alpha^2}{840}(2 + 12\beta + 7\beta^2) + \frac{\alpha}{840}(1 + 6\beta). \tag{3.88}$$

Table 3.2 shows that, for all but the smallest values of α, the first improvement given by (3.87) is superior as a lower bound to the second improvement given by (3.88). The bounds on the motion (3.61), (3.70) and (3.84) are shown graphically in Fig. 3.2.

3.6.2 Cylindrical freezing. For cylindrical freezing we utilize

$$-\frac{1}{\dot{R}(t)} \geq \alpha R(\beta - \log R), \tag{3.89}$$

which follows from (3.26). From (3.25) and (3.27) we have

$$t = t_{pss}(R) + \frac{d}{dt}\int_R^1 \int_R^\xi \xi\eta(\beta - \log \xi)(\log \xi - \log \eta)[\alpha + T(\eta, t)]\, d\eta\, d\xi, \tag{3.90}$$

where $t_{pss}(R)$ is defined by (1.74). On integrating this equation and using the inequalities (3.89) and $T \geq 0$ in the usual way we obtain finally

$$t^2 \geq \frac{\alpha^2}{16}[(1 + 2\beta)(1 - R^2) + 2R^2 \log R]^2$$

$$+ \frac{\alpha}{32}[(1 - R^2)(1 + 4\beta + (5 + 4\beta)R^2) + 4R^2(2 + 4\beta + R^2)\log R]. \tag{3.91}$$

Similarly, a further integration yields, after a long calculation,

$$t^3 \geq \frac{\alpha^3}{64}[(1 + 2\beta)(1 - R^2) + 2R^2 \log R]^3$$

$$+ \frac{\alpha^2}{1152}\Big\{(R^2 - 1)[(72\beta^2 + 114\beta + 19)R^4 + (180\beta^2 + 114\beta + 19)R^2$$

$$- 4(9\beta^2 + 12\beta + 2)] - 6[(19 + 24\beta)R^4 + 72\beta(1 + \beta)R^2$$

$$- 9(1 + 4\beta)]R^2 \log R + 72[3(1 + 2\beta) + R^2]R^4(\log R)^2\Big\}$$

$$+ \frac{\alpha}{384}\Big\{(1 - R^2)(10R^4 + 19R^2 + 1) + 6(R^4 + 6R^2 + 3)R^2 \log R$$

$$+ 6\beta[(1 - R^2)(R^4 + 10R^2 + 1) + 12(R^2 + 1)R^2 \log R]\Big\}. \tag{3.92}$$

In particular, for t_c we have from (3.91) and (3.92) the inequalities

$$t_c^2 \geq \frac{\alpha^2}{16}(1 + 2\beta)^2 + \frac{\alpha}{32}(1 + 4\beta), \tag{3.93}$$

$$t_c^3 \geq \frac{\alpha^3}{64}(1 + 2\beta)^3 + \frac{\alpha^2}{288}(2 + 12\beta + 9\beta^2) + \frac{\alpha}{384}(1 + 6\beta) \tag{3.94}$$

and again Table 3.3 shows that the first improvement (3.93) is a superior lower bound for all but the smallest α. The various bounds on the motion (3.63), (3.72) and (3.91) are shown in Fig. 3.3.

3.7 Summary

Most of the results of this chapter can be summarized in terms of the general equation (1.34), where $\lambda = 0$, 1 and 2 for the slab $(-\infty, 1)$, cylinder and sphere geometries, respectively, and where for the slab the position x and boundary position $X(t)$ are related to r and $R(t)$, respectively, by (1.35). With this terminology the time t_c for complete solidification for the slab is simply the time taken for the boundary to move a unit distance. We introduce the function $K_\lambda(x, y)$ defined by

$$K_\lambda(x, y) = \int_y^x \xi^{-\lambda} \, d\xi. \tag{3.95}$$

In this notation the pseudo-steady-state approximation becomes

$$T_{pss}(r, t) = \frac{K_\lambda(r, R)}{\beta + K_\lambda(1, R)}, \tag{3.96}$$

$$t_{pss}(R) = \alpha \int_R^1 \xi^\lambda [\beta + K_\lambda(1, \xi)] \, d\xi. \tag{3.97}$$

Moreover, with $u(r, t)$ defined by (3.44), the Green's function for the self-adjoint problem

$$\frac{\partial}{\partial r}\left(r^\lambda \frac{\partial u}{\partial r}\right) = r^\lambda \frac{\partial T}{\partial t}, \qquad R(t) < r < 1, \tag{3.98}$$

with boundary conditions (3.46) becomes

$$\begin{aligned} G(r, \xi, t) &= -\frac{K_\lambda(\xi, R)[\beta + K_\lambda(1, r)]}{[\beta + K_\lambda(1, R)]}, \qquad R \leq \xi \leq r, \\ &= -\frac{K_\lambda(r, R)[\beta + K_\lambda(1, \xi)]}{[\beta + K_\lambda(1, R)]}, \qquad r \leq \xi \leq 1, \end{aligned} \tag{3.99}$$

and we have

$$T(r, t) = T_{pss}(r, t) + \int_R^1 G(r, \xi, t)\xi^\lambda \frac{\partial T}{\partial t}(\xi, t)\, d\xi. \tag{3.100}$$

On differentiating this equation with respect to r and setting r to $R(t)$ we obtain from the Stefan condition $(2.6)_1$

$$1 = -\alpha R^\lambda[\beta + K_\lambda(1, R)]\dot{R}(t) + \int_R^1 \xi^\lambda[\beta + K_\lambda(1, \xi)]\frac{\partial T}{\partial t}(\xi, t)\, d\xi, \tag{3.101}$$

which may be integrated to give

$$t = \int_R^1 \xi^\lambda[\beta + K_\lambda(1, \xi)][\alpha + T(\xi, t)]\, d\xi, \tag{3.102}$$

and this is the appropriate generalization of Eqs. (3.6), (3.23) and (3.27). On multiplying T_{pss} in (3.100) by the right-hand side of (3.101) (which is unity) we obtain, on using (3.96) and simplifying the result,

$$T(r, t) = \frac{\partial}{\partial t}\int_R^r \xi^\lambda K_\lambda(r, \xi)[\alpha + T(\xi, t)]\, d\xi, \tag{3.103}$$

which is the appropriate generalization (3.4), (3.21) and (3.25).

From (3.102) the bounds resulting from $0 \le T \le 1$ are

$$\alpha \int_R^1 \xi^\lambda[\beta + K_\lambda(1, \xi)]\, d\xi \le t \le (\alpha + 1)\int_R^1 \xi^\lambda[\beta + K_\lambda(1, \xi)]\, d\xi, \tag{3.104}$$

while the generalization of the improved upper bound resulting from $T \le T_{pss}$ becomes

$$t \le \alpha \int_R^1 \xi^\lambda[\beta + K_\lambda(1, \xi)]\, d\xi + \int_R^1 \xi^\lambda \frac{[\beta + K_\lambda(1, \xi)]K_\lambda(\xi, R)}{[\beta + K_\lambda(1, R)]}\, d\xi. \tag{3.105}$$

For the improved lower bounds the general inequality

$$-\frac{1}{\dot{R}(t)} \ge \alpha R^\lambda[\beta + K_\lambda(1, R)] = -\frac{dt_{pss}}{dR}(R) \tag{3.106}$$

follows immediately from (3.101), noting $\partial T/\partial t \ge 0$ and (3.97). Using (3.106) and $T \ge 0$ the improved lower bounds arising from integrating

(3.102) become

$$t^2 \geq t^2_{\text{pss}}(R) + 2\alpha \int_R^1 \int_R^\xi (\xi\eta)^\lambda [\beta + K_\lambda(1, \xi)]K_\lambda(\xi, \eta) \, d\eta \, d\xi, \qquad (3.107)$$

$$t^3 \geq t^3_{\text{pss}}(R)$$

$$+ 6\alpha^2 \int_R^1 \int_\zeta^1 \int_\zeta^\xi (\xi\eta\zeta)^\lambda [\beta + K_\lambda(1, \xi)]K_\lambda(\xi, \eta)[\beta + K_\lambda(1, \zeta)] \, d\eta \, d\xi \, d\zeta$$

$$+ 6\alpha \int_R^1 \int_R^\xi \int_R^\eta (\xi\eta\zeta)^\lambda [\beta + K_\lambda(1, \xi)]K_\lambda(\xi, \eta)K_\lambda(\eta, \zeta) \, d\zeta \, d\eta \, d\xi, \qquad (3.108)$$

which agree with those given in the previous section.

The various bounds for the time t_c to complete solidification may be summarized compactly in terms of the parameter λ. The bounds resulting from (3.104) yield

$$\frac{\alpha(1 + 2\beta)}{2(\lambda + 1)} \leq t_c \leq \frac{(\alpha + 1)(1 + 2\beta)}{2(\lambda + 1)}, \qquad (3.109)$$

and it is of some interest to note that although for the sphere and cylinder the improved upper bound (3.105) coincides with that from (3.104), for the slab, however, (3.105) gives

$$t_c \leq \frac{\alpha}{2}(1 + 2\beta) + \frac{1 + 3\beta}{6(1 + \beta)}, \qquad (3.110)$$

which is a genuine improvement on the upper bound in (3.109) with $\lambda = 0$. We note that the result (3.110) is apparent from (3.66) on setting $X = 1$. For the improved lower bounds for t_c we obtain from (3.107) and (3.108), respectively,

$$t_c^2 \geq \left(\frac{\alpha(1 + 2\beta)}{2(\lambda + 1)}\right)^2 + \frac{\alpha(1 + 4\beta)}{4(\lambda + 1)(\lambda + 3)}, \qquad (3.111)$$

$$t_c^3 \geq \left(\frac{\alpha(1 + 2\beta)}{2(\lambda + 1)}\right)^3 + \frac{\alpha^2(\lambda^2 + 5\lambda + 10)}{8(\lambda + 1)^2(\lambda + 2)(\lambda + 3)(\lambda + 5)}$$

$$\times \left\{1 + 6\beta + \frac{12\beta^2(\lambda + 5)}{\lambda^2 + 5\lambda + 10}\right\} + \frac{\alpha(1 + 6\beta)}{8(\lambda + 1)(\lambda + 3)(\lambda + 5)}. \qquad (3.112)$$

We remark that although the first expression (3.111) is apparent from special cases, the latter is by no means obvious. After a long calculation the above result (3.112) can be established directly from (3.95) and (3.108) with λ arbitrary.

Finally in this chapter we comment that the essential problem remaining to determine tighter bounds is to obtain a lower bound on the velocity

$-\dot{R}(t)$ of the boundary. If a non-trivial bound were known we might utilize

$$T(r, t) \geq -\alpha \dot{R}(t) R^\lambda K_\lambda(r, R), \tag{3.113}$$

which follows from (3.103) to improve both the lower and upper bounds. From (3.102) and (3.113) we would immediately have a new lower bound

$$t \geq \alpha \int_R^1 \xi^\lambda [\beta + K_\lambda(1, \xi)][1 - \dot{R}(t) R^\lambda K_\lambda(\xi, R)] \, d\xi. \tag{3.114}$$

Moreover, from (3.103) we have on integrating (3.102) as in the previous section

$$\frac{t^2}{2} = \int_R^1 \frac{t_{\text{pss}}(R(v))}{-\dot{R}(v)} \, dv$$
$$+ \int_R^1 \int_R^\xi (\eta\xi)^\lambda [\beta + K_\lambda(1, \xi)] K_\lambda(\xi, \eta)[\alpha + T(\eta, t)] \, d\eta \, d\xi, \tag{3.115}$$

and this equation together with either $T \leq 1$ or $T \leq T_{\text{pss}}$ and a lower bound on $-\dot{R}(t)$ would give rise to new upper bounds. For example, for the exact Neumann solution $X(t) = (2\gamma t)^{1/2}$ we have from (1.59)

$$\frac{1}{(\alpha + \frac{1}{3})X} \leq \dot{X}(t) \leq \frac{1}{\alpha X}, \tag{3.116}$$

and using this lower bound for $\dot{X}(t)$, the equivalent equation to (3.114) gives

$$t \geq \frac{\alpha}{2} \left(\frac{2 + 3\alpha}{1 + 3\alpha} \right) X^2. \tag{3.117}$$

Further, from the equivalent equation to (3.115) and $T \leq 1$ and $T \leq T_{\text{pss}}$ we obtain, respectively,

$$t^2 \leq \tfrac{1}{12}(1 + 2\alpha + 3\alpha^2) X^4, \tag{3.118}$$

$$t^2 \leq \tfrac{1}{60}(1 + 10\alpha + 15\alpha^2) X^4, \tag{3.119}$$

the latter being the tighter inequality. From the above the following inequalities for γ are apparent:

$$\frac{1}{\gamma} \geq \alpha \left(\frac{2 + 3\alpha}{1 + 3\alpha} \right), \tag{3.120}$$

$$\frac{1}{\gamma^2} \leq \tfrac{1}{15}(1 + 10\alpha + 15\alpha^2) \leq \tfrac{1}{3}(1 + 2\alpha + 3\alpha^2), \tag{3.121}$$

which may be verified numerically. Clearly, we might now utilize (3.120) and (3.121) to deduce further inequalities for γ in precisely the same way. We do not pursue this matter further and emphasize only that for the general problem with β and λ non-zero, important results hinge on the determination of a non-trivial lower bound for the velocity $-\dot{R}(t)$.

3.8   Additional symbols used

a, b	constants $[a, b]$ is the interval of a real line, (3.28)
$f(x), p(x), q(x)$	arbitrary functions such that $p(x)$ is non-zero in interval $[a, b]$, (3.28)
G	Green's function, (3.31)
K_λ	function defined by (3.95)
t_ξ	function of ξ defined by $\xi = R(t_\xi)$, (3.54)
u	difference between temperature and pseudo-steady-state estimate $T - T_{\text{pss}}$, (3.34)
u_1, u_2	linearly independent solutions of a homogeneous equation, (3.37)
w_0	constant involved in Wronskian of $y_1(x)$ and $y_2(x)$, (3.33)
$y(x)$	solution of (3.28) and (3.29)
$y_1(x), y_2(x)$	linearly independent solutions of homogeneous equation (3.28)

Greek symbols

ζ, η, ν, ω	integation variables
λ_1, λ_2	arbitrary constants, (3.29)
$\phi(x, \xi, t)$	arbitrary function such that $\phi(x, X, t)$ is non-zero, (3.13)

Convention

Dot denotes differentiation with respect to time t.

4

Integral equations

4.1 Introduction

In this chapter we utilize the integral formulation of the previous chapter to deduce integral equations for the three basic problems. The first integral equations are developed via formal series solutions while the second follow immediately from the integral formulation on eliminating the variable time, so as to obtain a single equation for the temperature as a function of position and boundary position. These latter equations are widely used as integral iteration schemes starting with the pseudo-steady-state estimate as the initial approximation.

In the following section we deduce formal series solutions by repeated application of the basic integral formulation. These series provide exact expressions for the temperature assuming the boundary position is a known function. Since this is not known, these results are of limited practical use. However, they do provide some indication of the nature of the mathematical complexities. For the cylinder the series solution is far more complicated than those for the plane and sphere and involves the functions $c_n(z, z_0)$ and $e_n(z, z_0)$ first introduced by Langford (1967b). The first four of these functions are given by (2.4) and (2.5), respectively, while general expressions are derived in Appendix 1. With

$$c_0(z, z_0) = 1, \quad e_0(z, z_0) = \log\left(\frac{z}{z_0}\right), \tag{4.1}$$

these functions are generated by the same integral formulae (see Appendix 1)

$$c_n(z, z_0) = \int_{z_0}^{z} \log\left(\frac{z}{\omega}\right) c_{n-1}(\omega, z_0) \, d\omega \qquad (n \geq 1), \tag{4.2}$$

$$e_n(z, z_0) = \int_{z_0}^{z} \log\left(\frac{z}{\omega}\right) e_{n-1}(\omega, z_0) \, d\omega \qquad (n \geq 1), \tag{4.3}$$

82

and, moreover, are connected by

$$c_n(z, z_0) = \int_{z_0}^{z} e_{n-1}(z, \omega)\, \mathrm{d}\omega \qquad (n \geq 1).\tag{4.4}$$

Clearly, from $(4.1)_2$ and (4.3) we have

$$e_n(z, z_0) = \int_{z_0}^{z} e_0(z, \omega)e_{n-1}(\omega, z_0)\, \mathrm{d}\omega \qquad (n \geq 1),\tag{4.5}$$

and this equation turns out to be precisely that needed in verifying the formal series solution for the cylinder. In Section 4.3 we utilize the formal series solutions to deduce unsolved integral equations for the boundary motion. These equations are rather unusual and the only simple observation we may deduce from them is that the pseudo-steady-state motion approximates the short-time boundary motion. For plane solidification Grinberg and Chekmareva (1971) employ asymptotic methods to estimate the large-time solution. Since their analysis involves a number of assumptions and the results obtained are not particularly explicit we refer the reader to their original paper for further details. The unsolved integral equations given in Section 4.3 can be alternatively derived by taking a 'modified' Laplace transform of the formal series solution for the temperature.

In Section 4.4 we consider an integral iteration scheme which is frequently employed to obtain approximate analytic estimates. For each of the three problems we give one iteration of the pseudo-steady-state temperature and the corresponding motion of the boundary. These estimates can be shown to satisfy known bounds. For the plane problem the approximate motion of the boundary is compared with the nine-term perturbation solution (2.78) due to Pedroso and Domoto (1973a) and the agreement is found to be very close indeed. We may therefore infer that the approximate motion for the plane problem (4.68) is at least valid for $\alpha \geq 2$. For the spherical and cylindrical problems, although the approximate motions are reasonably well behaved, theoretical and numerical results indicate certain undesirable features for times close to complete freezing. In the final section of the chapter results for the integral iteration scheme are summarized in terms of the general equation (1.34) and the function $K_\lambda(x, y)$ defined by (3.95). In addition, we note an alternative iterative scheme from which both the large α expansions of Chapter 2 and the integral iterations of Section 4.4 can be seen to emerge.

4.2 Formal series solutions

In this section we merely indicate the crucial steps in the derivation and we make no attempt to provide a completely rigorous justification of all the various formal manipulations. For plane solidification we may write (3.4) as

$$T(x, t) = \alpha \frac{\partial}{\partial t} \frac{(X - x)^2}{2!} + \frac{\partial}{\partial t} \int_x^X (\eta - x)T(\eta, t)\, d\eta, \tag{4.6}$$

and we now use (3.4) again to obtain an expression for the integral in (4.6). We have

$$\int_x^X (\eta - x)T(\eta, t)\, d\eta = \int_x^X (\eta - x) \frac{\partial}{\partial t} \int_\eta^X (\xi - \eta)[\alpha + T(\xi, t)]\, d\xi\, d\eta$$

$$= \frac{\partial}{\partial t} \int_x^X \int_\eta^X (\xi - \eta)(\eta - x)[\alpha + T(\xi, t)]\, d\xi\, d\eta$$

$$= \frac{\partial}{\partial t} \int_x^X \left(\int_x^\xi (\xi - \eta)(\eta - x)\, d\eta \right)[\alpha + T(\xi, t)]\, d\xi$$

$$= \frac{\partial}{\partial t} \int_x^X \frac{(\xi - x)^3}{3!}[\alpha + T(\xi, t)]\, d\xi, \tag{4.7}$$

which on substituting into (4.6) gives

$$T(x, t) = \alpha \frac{\partial}{\partial t} \frac{(X - x)^2}{2!} + \alpha \frac{\partial^2}{\partial t^2} \frac{(X - x)^4}{4!} + \frac{\partial^2}{\partial t^2} \int_x^X \frac{(\xi - x)^3}{3!} T(\xi, t)\, d\xi. \tag{4.8}$$

Proceeding in this manner we may evidently deduce that, for any integer $N \geq 1$,

$$T(x, t) = \alpha \sum_{n=1}^N \frac{\partial^n}{\partial t^n} \frac{(X - x)^{2n}}{(2n)!} + \frac{\partial^N}{\partial t^N} \int_x^X \frac{(\xi - x)^{2N-1}}{(2N - 1)!} T(\xi, t)\, d\xi, \tag{4.9}$$

and thus assuming the remainder tends to zero as N tends to infinity we have

$$T(x, t) = \alpha \sum_{n=1}^\infty \frac{\partial^n}{\partial t^n} \frac{(X - x)^{2n}}{(2n)!}. \tag{4.10}$$

We may verify directly that (4.10) represents a formal series solution of (2.1), (2.2)$_2$ and (2.3)$_1$. For the right-hand side of (2.1) we have from (4.10)

$$\frac{\partial^2 T}{\partial x^2} = \alpha \sum_{n=2}^\infty \frac{\partial^n}{\partial t^n} \frac{(X - x)^{2n-2}}{(2n - 2)!} = \alpha \sum_{m=1}^\infty \frac{\partial^{m+1}}{\partial t^{m+1}} \frac{(X - x)^{2m}}{(2m)!}, \tag{4.11}$$

on setting $m = n - 1$. This expression is clearly the time partial derivative of (4.10) and therefore (4.10) satisfies (2.1). Moreover, it is immediately apparent that (4.10) and its partial derivative with respect to x have the appropriate values on the boundary. Alternatively, we may show that (4.10) satisfies the basic integral formulation (3.4) as follows:

$$\frac{\partial}{\partial t} \int_x^X (\xi - x)[\alpha + T(\xi, t)] \, d\xi = \alpha \frac{\partial}{\partial t} \int_x^X (\xi - x) \sum_{n=0}^{\infty} \frac{\partial^n}{\partial t^n} \frac{(X - \xi)^{2n}}{(2n)!} \, d\xi$$

$$= \alpha \frac{\partial}{\partial t} \sum_{n=0}^{\infty} \int_x^X (\xi - x) \frac{\partial^n}{\partial t^n} \frac{(X - \xi)^{2n}}{(2n)!} \, d\xi$$

$$= \alpha \sum_{n=0}^{\infty} \frac{\partial^{n+1}}{\partial t^{n+1}} \int_x^X \frac{(X - \xi)^{2n}}{(2n)!} (\xi - x) \, d\xi$$

$$= \alpha \sum_{n=0}^{\infty} \frac{\partial^{n+1}}{\partial t^{n+1}} \frac{(X - x)^{2n+2}}{(2n + 2)!}, \qquad (4.12)$$

and $m = n + 1$ gives precisely (4.10).

Now from (3.6) and (4.10) we have

$$t = \alpha \int_0^X (\xi + \beta) \sum_{n=0}^{\infty} \frac{\partial^n}{\partial t^n} \frac{(X - \xi)^{2n}}{(2n)!} \, d\xi, \qquad (4.13)$$

from which we may readily deduce

$$\frac{t}{\alpha} = \sum_{n=0}^{\infty} \frac{\partial^n}{\partial t^n} \left\{ \frac{X^{2n+2}}{(2n + 2)!} + \beta \frac{X^{2n+1}}{(2n + 1)!} \right\}. \qquad (4.14)$$

We note that this equation may also be derived directly from an integration of $(2.2)_1$ after using (4.10). From the exact Neumann solution $X(t) = (2\gamma t)^{1/2}$ for β zero we may readily verify that the series (1.56) emerges from (4.14), since for $n \geq 0$ we have

$$\frac{\partial^n}{\partial t^n} X^{2n+2} = (2\gamma)^{n+1} \frac{\partial^n}{\partial t^n} t^{n+1} = (n + 1)! \, (2\gamma)^{n+1} t. \qquad (4.15)$$

Equation (4.14) is the starting point for the first derivation of the unsolved integral equation for $X(t)$ given in the following section. We close this section with a summary of the corresponding results for spheres and cylinders.

4.2.1 Spherical freezing.

From (3.21) we may show as described above that

$$T(r, t) = \frac{\alpha}{r} \frac{\partial}{\partial t} \int_R^r \xi(r - \xi) \, d\xi + \frac{\alpha}{r} \frac{\partial^2}{\partial t^2} \int_R^r \xi \frac{(r - \xi)^3}{3!} \, d\xi$$

$$+ \frac{1}{r} \frac{\partial^2}{\partial t^2} \int_R^r \xi \frac{(r - \xi)^3}{3!} T(\xi, t) \, d\xi. \qquad (4.16)$$

Continuing this process and assuming the remainder tends to zero we obtain

$$T(r, t) = \frac{\alpha}{r} \sum_{n=1}^{\infty} \frac{\partial^n}{\partial t^n} \int_R^r \xi \frac{(r - \xi)^{2n-1}}{(2n - 1)!} \, d\xi, \tag{4.17}$$

and this equation simplifies to give

$$T(r, t) = \frac{\alpha}{r} \sum_{n=1}^{\infty} \frac{\partial^n}{\partial t^n} \left\{ \frac{(r - R)^{2n}}{(2n + 1)!} (r + 2nR) \right\}, \tag{4.18}$$

as the formal solution of (2.4), (2.5)$_2$ and (2.6)$_1$. Clearly, (4.18) satisfies both conditions on the boundary. Further, we may show

$$\frac{\partial^2 T}{\partial r^2} + \frac{2}{r} \frac{\partial T}{\partial r} = \frac{\alpha}{r} \sum_{n=2}^{\infty} \frac{\partial^n}{\partial t^n} \left\{ \frac{(r - R)^{2n-2}}{(2n - 1)!} [r + 2(n - 1)R] \right\}, \tag{4.19}$$

and on setting $m = n - 1$ the series in (4.19) is simply the time derivative of (4.18) and therefore (4.18) satisfies (2.4). Alternatively, we may verify (4.18) satisfies (3.21) as follows:

$$\frac{1}{r} \frac{\partial}{\partial t} \int_R^r \xi(r - \xi)[\alpha + T(\xi, t)] \, d\xi$$

$$= \frac{\alpha}{r} \frac{\partial}{\partial t} \int_R^r (r - \xi) \sum_{n=0}^{\infty} \frac{\partial^n}{\partial t^n} \left\{ \frac{(\xi - R)^{2n}}{(2n + 1)!} (\xi + 2nR) \right\} d\xi$$

$$= \frac{\alpha}{r} \frac{\partial}{\partial t} \sum_{n=0}^{\infty} \int_R^r (r - \xi) \frac{\partial^n}{\partial t^n} \left\{ \frac{(\xi - R)^{2n}}{(2n + 1)!} (\xi + 2nR) \right\} d\xi$$

$$= \frac{\alpha}{r} \sum_{n=0}^{\infty} \frac{\partial^{n+1}}{\partial t^{n+1}} \left\{ \int_R^r \frac{(\xi - R)^{2n}}{(2n + 1)!} (\xi + 2nR)(r - \xi) \, d\xi \right\}$$

$$= \frac{\alpha}{r} \sum_{n=0}^{\infty} \frac{\partial^{n+1}}{\partial t^{n+1}} \left\{ \frac{(r - R)^{2n+2}}{(2n + 3)!} [r + 2(n + 1)R] \right\}, \tag{4.20}$$

and $m = n + 1$ gives the desired result. From (4.18) and (3.23) we may deduce

$$\frac{t}{\alpha} = \sum_{n=0}^{\infty} \frac{\partial^n}{\partial t^n} \left\{ \frac{(1 - R)^{2n+2}}{(2n + 3)!} [1 + 2(n + 1)R] \right.$$

$$\left. + \frac{\beta(1 - R)^{2n+1}}{(2n + 1)!} \frac{[1 + (2n + 1)R + R^2]}{(2n + 3)} \right\}, \tag{4.21}$$

which again can also be deduced directly from (2.5)$_1$ and (4.18).

4.2.2 Cylindrical freezing. From two applications of (3.25) we have,

as previously described,

$$T(r, t) = \alpha \frac{\partial}{\partial t} \int_R^r \xi (\log r - \log \xi) \, d\xi$$

$$+ \frac{\partial^2}{\partial t^2} \int_R^r \frac{\xi}{4} \left\{ (\xi^2 - r^2) + (\xi^2 + r^2)(\log r - \log \xi) \right\} [\alpha + T(\xi, t)] \, d\xi. \tag{4.22}$$

Now on making the transformations

$$z = \frac{r^2}{4}, \qquad Z(t) = \frac{R(t)^2}{4}, \qquad \omega = \frac{\xi^2}{4}, \tag{4.23}$$

we have from (2.25) that (4.22) becomes

$$T(r, t) = \alpha \frac{\partial}{\partial t} \int_Z^z e_0(z, \omega) \, d\omega + \alpha \frac{\partial^2}{\partial t^2} \int_Z^z e_1(z, \omega) \, d\omega$$

$$+ \frac{\partial^2}{\partial t^2} \int_Z^z e_1(z, \omega) T(2\omega^{1/2}, t) \, d\omega. \tag{4.24}$$

As before we are led to consider

$$T(r, t) = \alpha \sum_{n=1}^{\infty} \frac{\partial^n}{\partial t^n} \int_Z^z e_{n-1}(z, \omega) \, d\omega, \tag{4.25}$$

as a formal solution of (2.7), (2.5)$_2$ and (2.6)$_1$. The series satisfies both conditions on the boundary, and to verify (2.7) is satisfied we have

$$\frac{\partial^2 T}{\partial r^2} + \frac{1}{r} \frac{\partial T}{\partial r} = \alpha \sum_{n=2}^{\infty} \frac{\partial^n}{\partial t^n} \int_Z^z \frac{\partial}{\partial z} \left(z \frac{\partial e_{n-1}}{\partial z} (z, \omega) \right) d\omega, \tag{4.26}$$

and from (A1.1) together with $m = n - 1$ we have the desired result. Alternatively, we may verify that (4.25) satisfies (3.25) as follows. We have

$$\int_R^r \xi (\log r - \log \xi) T(\xi, t) \, d\xi$$

$$= \int_Z^z e_0(z, \omega) T(2\omega^{1/2}, t) \, d\omega$$

$$= \alpha \int_Z^z e_0(z, \omega) \sum_{n=1}^{\infty} \frac{\partial^n}{\partial t^n} \int_Z^{\omega} e_{n-1}(\omega, \Omega) \, d\Omega \, d\omega$$

$$= \alpha \sum_{n=1}^{\infty} \frac{\partial^n}{\partial t^n} \int_Z^z \int_Z^{\omega} e_0(z, \omega) e_{n-1}(\omega, \Omega) \, d\Omega \, d\omega$$

$$= \alpha \sum_{n=1}^{\infty} \frac{\partial^n}{\partial t^n} \int_Z^z \int_{\Omega}^z e_0(z, \omega) e_{n-1}(\omega, \Omega) \, d\omega \, d\Omega$$

$$= \alpha \sum_{n=1}^{\infty} \frac{\partial^n}{\partial t^n} \int_Z^z e_n(z, \Omega) \, d\Omega, \tag{4.27}$$

where we have used (4.5). Thus, altogether the right-hand side of (3.25) becomes

$$\alpha \frac{\partial}{\partial t} \int_Z^z e_0(z, \omega) \, d\omega + \alpha \sum_{n=1}^\infty \frac{\partial^{n+1}}{\partial t^{n+1}} \int_Z^z e_n(z, \omega) \, d\omega, \tag{4.28}$$

which on simplification yields precisely the series (4.25).

From (4.4), (4.23) and (4.25) the formal solution for cylindrical freezing becomes

$$T(r, t) = \alpha \sum_{n=1}^\infty \frac{\partial^n}{\partial t^n} c_n\left(\frac{r^2}{4}, \frac{R(t)^2}{4}\right), \tag{4.29}$$

and from this equation and $(2.5)_1$ the formal equation for the motion of the boundary becomes

$$\frac{t}{\alpha} = \sum_{n=1}^\infty \frac{\partial^{n-1}}{\partial t^{n-1}} \left\{ c_n\left(\frac{1}{4}, \frac{R(t)^2}{4}\right) + \frac{\beta}{2} \frac{\partial c_n}{\partial z}\left(\frac{1}{4}, \frac{R(t)^2}{4}\right) \right\}. \tag{4.30}$$

4.3 Unsolved integral equations for boundary motion

In this section we deduce integral equations for the motion of the boundary. We make use of the result that if $f(t)$ is any function such that

$$f(0) = f'(0) = \ldots = f^{(n-1)}(0) = 0, \tag{4.31}$$

then

$$\int_0^\infty e^{-pt} f^{(n)}(t) \, dt = p^n \int_0^\infty e^{-pt} f(t) \, dt, \tag{4.32}$$

where $f^{(n)}(t)$ denotes the nth derivative. Noting this result and taking the Laplace transform of (4.14) gives

$$\frac{1}{\alpha p^2} = \sum_{n=0}^\infty \int_0^\infty e^{-pt} p^n \left\{ \frac{X(t)^{2n+2}}{(2n+2)!} + \frac{\beta X(t)^{2n+1}}{(2n+1)!} \right\} dt, \tag{4.33}$$

which we may rearrange to obtain

$$\int_0^\infty e^{-pt} \left\{ \cosh p^{1/2} X(t) + \beta p^{1/2} \sinh p^{1/2} X(t) \right\} dt = \left(1 + \frac{1}{\alpha}\right) \frac{1}{p}. \tag{4.34}$$

This is the basic unsolved integral equation for $X(t)$ for planar freezing. We make the following observations. First, for β zero we may readily confirm the Neumann solution $X(t) = (2\gamma t)^{1/2}$, since the left-hand side of

(4.34) becomes

$$\frac{1}{2}\int_0^\infty e^{-pt}\left\{e^{(2\gamma pt)^{1/2}} + e^{-(2\gamma pt)^{1/2}}\right\} dt$$

$$= \frac{e^{\gamma/2}}{2}\int_0^\infty \left\{e^{-[(pt)^{1/2}-(\gamma/2)^{1/2}]^2} + e^{-[(pt)^{1/2}+(\gamma/2)^{1/2}]^2}\right\} dt$$

$$= \frac{e^{\gamma/2}}{p}\left\{\int_{-(\gamma/2)^{1/2}}^\infty e^{-u^2}[u + (\gamma/2)^{1/2}] \, du + \int_{(\gamma/2)^{1/2}}^\infty e^{-u^2}[u - (\gamma/2)^{1/2}] \, du\right\},$$

$$(4.35)$$

which simplifies so that together with the right-hand side of (4.34) we obtain the transcendental equation (1.43) for γ. Secondly, for small $X(t)$ we have formally from (4.34), by expanding cosh and sinh,

$$\int_0^\infty e^{-pt}\left\{\frac{X(t)^2}{2} + \beta X(t)\right\} dt \simeq \frac{1}{\alpha p^2}, \tag{4.36}$$

and thus

$$\frac{X(t)^2}{2} + \beta X(t) \simeq \frac{t}{\alpha}, \tag{4.37}$$

which is simply the pseudo-steady-state motion. Thirdly, an alternative equation for the inverse motion $X^{-1}(x)$ may be deduced from (4.34) by integration by parts. On integrating the exponential we have

$$\int_0^\infty e^{-pt}\left\{\frac{\sinh p^{1/2}X(t)}{p^{1/2}} + \beta \cosh p^{1/2}X(t)\right\}\frac{dX(t)}{dt} \, dt = \frac{1}{\alpha p}, \tag{4.38}$$

and therefore

$$\int_0^\infty e^{-pX^{-1}(\xi)}\left\{\frac{\sinh p^{1/2}\xi}{p^{1/2}} + \beta \cosh p^{1/2}\xi\right\} d\xi = \frac{1}{\alpha p}, \tag{4.39}$$

which is the form usually adopted in the literature (see, for example, Ockendon 1975).

We may also derive the above equations by means of modified Laplace transforms. In taking the Laplace transform of (4.10) we observe that $[X(t) - x]^{2n}$, and appropriate time derivatives are zero on the boundary $x = X(t)$ and therefore we use the transform

$$\hat{T}(x, p) = \int_{X^{-1}(x)}^\infty e^{-pt}T(x, t) \, dt. \tag{4.40}$$

From (4.10), (4.40) and the corresponding result to (4.32) we readily obtain

$$\hat{T}(x, p) = \alpha \int_{X^{-1}(x)}^{\infty} e^{-pt} \left\{ \cosh p^{1/2}[X(t) - x] - 1 \right\} dt, \tag{4.41}$$

and by integration by parts we have

$$\hat{T}(x, p) = \alpha \int_{X^{-1}(x)}^{\infty} e^{-pt} \frac{\sinh p^{1/2}[X(t) - x]}{p^{1/2}} \frac{dX(t)}{dt} dt. \tag{4.42}$$

Thus, with the change of variable $\xi = X(t)$ we may deduce

$$\hat{T}(x, p) = \alpha \int_{x}^{\infty} e^{-pX^{-1}(\xi)} \frac{\sinh p^{1/2}(\xi - x)}{p^{1/2}} d\xi, \tag{4.43}$$

and the three versions of the integral equation for the motion $X(t)$, namely (4.34), (4.38) and (4.39), can be alternatively deduced from (4.41), (4.42) and (4.43), respectively, and the surface condition $(2.2)_1$. For both spherical and cylindrical geometries we proceed via this latter approach.

4.3.1 Spherical freezing.

In this case we define the modified Laplace transform

$$\hat{T}(r, p) = \int_{R^{-1}(r)}^{\infty} e^{-pt} T(r, t) dt, \tag{4.44}$$

where $R^{-1}(r)$ denotes the inverse of $r = R(t)$ such that $R^{-1}(0) = \infty$. From (4.18) we have

$$\hat{T}(r, p) = \frac{\alpha}{r} \int_{R^{-1}(r)}^{\infty} \sum_{n=1}^{\infty} e^{-pt} p^n \frac{[r - R(t)]^{2n}}{(2n + 1)!} \left\{ [r - R(t)] + (2n + 1)R(t) \right\} dt, \tag{4.45}$$

which simplifies to give

$$\hat{T}(r, p) = \frac{\alpha}{r} \int_{R^{-1}(r)}^{\infty} e^{-pt} \left\{ \frac{\sinh p^{1/2}[r - R(t)]}{p^{1/2}} + R(t) \cosh p^{1/2}[r - R(t)] - r \right\} dt. \tag{4.46}$$

Integration by parts gives simply

$$\hat{T}(r, p) = -\frac{\alpha}{r} \int_{R^{-1}(r)}^{\infty} e^{-pt} R(t) \frac{\sinh p^{1/2}[r - R(t)]}{p^{1/2}} \frac{dR(t)}{dt} dt, \tag{4.47}$$

and therefore from the change of variable $\xi = R(t)$ we obtain

$$\hat{T}(r, p) = \frac{\alpha}{r} \int_0^r e^{-pR^{-1}(\xi)} \xi \frac{\sinh p^{1/2}(r - \xi)}{p^{1/2}} \, d\xi. \qquad (4.48)$$

From (4.46), (4.47) and (4.48) and the surface condition $(2.5)_1$ we obtain, respectively, the following versions of the integral equation for $R(t)$, thus

$$\int_0^\infty e^{-pt} \left\{ [(1 - \beta) + \beta p R(t)] \frac{\sinh p^{1/2}[1 - R(t)]}{p^{1/2}} \right.$$

$$\left. + [(1 - \beta)R(t) + \beta] \cosh p^{1/2}[1 - R(t)] - 1 \right\} dt = \frac{1}{\alpha p}, \quad (4.49)$$

$$\int_0^\infty e^{-pt} R(t) \left\{ (1 - \beta) \frac{\sinh p^{1/2}[1 - R(t)]}{p^{1/2}} + \beta \cosh p^{1/2}[1 - R(t)] \right\} \frac{dR(t)}{dt} \, dt$$

$$= -\frac{1}{\alpha p}, \quad (4.50)$$

$$\int_0^1 e^{-pR^{-1}(\xi)} \xi \left\{ (1 - \beta) \frac{\sinh p^{1/2}(1 - \xi)}{p^{1/2}} + \beta \cosh p^{1/2}(1 - \xi) \right\} d\xi = \frac{1}{\alpha p}.$$

$$(4.51)$$

Again by expanding cosh and sinh we may deduce from (4.49) that the pseudo-steady-state estimate (1.71) of the motion applies in the early stages of solidification.

4.3.2 Cylindrical freezing. From Appendix 1, Eqs. (A1.13) and (A1.15) we have

$$\sum_{n=0}^\infty c_n(z, z_0) p^n = 2(pz_0)^{1/2} \psi(2(pz_0)^{1/2}, 2(pz)^{1/2}), \qquad (4.52)$$

$$\sum_{n=0}^\infty e_n(z, z_0) p^n = 2\phi(2(pz)^{1/2}, 2(pz_0)^{1/2}), \qquad (4.53)$$

where $\phi(x, y)$ and $\psi(x, y)$ are defined by

$$\phi(x, y) = I_0(x) K_0(y) - I_0(y) K_0(x), \qquad (4.54)$$

$$\psi(x, y) = I_1(x) K_0(y) + I_0(y) K_1(x). \qquad (4.55)$$

Thus, with $\hat{T}(r, p)$ defined by (4.44) we have from (4.29)

$$\hat{T}(r, p) = \alpha \int_{R^{-1}(r)}^\infty e^{-pt} \left\{ 2(pZ)^{1/2} \psi(2(pZ)^{1/2}, 2(pz)^{1/2}) - 1 \right\} dt, \quad (4.56)$$

where z and $Z(t)$ are as given in Eq. (4.23). By integration by parts,

(4.56) gives

$$\hat{T}(r, p) = -\alpha \int_{R^{-1}(r)}^{\infty} e^{-pt} \phi(p^{1/2}r, p^{1/2}R(t))R(t) \frac{dR(t)}{dt} dt, \qquad (4.57)$$

and thus we have

$$\hat{T}(r, p) = \alpha \int_{0}^{r} e^{-pR^{-1}(\xi)} \xi \phi(p^{1/2}r, p^{1/2}\xi) \, d\xi. \qquad (4.58)$$

From the above representations and the surface condition $(2.5)_1$ we may deduce various forms of the integral equation for the motion of the boundary. For example, using (4.58) we obtain

$$\int_{0}^{1} e^{-pR^{-1}(\xi)} \xi \left\{ \phi(p^{1/2}, p^{1/2}\xi) + \beta p^{1/2} \psi(p^{1/2}, p^{1/2}\xi) \right\} d\xi = \frac{1}{\alpha p}. \qquad (4.59)$$

4.4 Integral iteration

Integral iteration of the pseudo-steady-state solution is frequently employed to obtain analytical solutions which hopefully converge to the exact solution. For example, for planar geometry Siegel and Savino (1966) and Savino and Siegel (1969) use this approach for determining the thickness of a frozen layer that forms when a warm liquid flows over a flat plate, cooled below the freezing temperature of the liquid by a coolant flowing along the other side of the plate. For spherical geometry Theofanous and Lim (1971) and Shih and Chou (1971) treat the spherical freezing problem for β zero and non-zero, respectively. Shih and Tsay (1971) consider cylindrical freezing with β non-zero. Moreover, these papers contain references to earlier applications of the integral iteration method.

For planar freezing we may formulate from (3.17) the following integral iteration procedure,

$$T_{n+1}(x, X) = \frac{\dfrac{\partial}{\partial X}\left(\displaystyle\int_{x}^{X} (\xi - x)[\alpha + T_n(\xi, X)] \, d\xi \right)}{\dfrac{\partial}{\partial X}\left(\displaystyle\int_{0}^{X} (\xi + \beta)[\alpha + T_n(\xi, X)] \, d\xi \right)}, \qquad (4.60)$$

for successive estimates $T_n(x, X)$ of the temperature $T^{\dagger}(x, X)$. Moreover, for each estimate of the temperature $T_n(x, X)$, we may deduce from (3.6) an estimate of the motion of the boundary from the equation

$$t_{n+1}(X) = \int_{0}^{X} (\xi + \beta)[\alpha + T_n(\xi, X)] \, d\xi. \qquad (4.61)$$

In principle, we may continue the above process indefinitely. However, although the actual calculations are straightforward the expressions obtained rapidly become lengthy and difficult to manipulate. The following two integrals which may be readily verified by integration by parts are useful:

$$\int_x^X (\xi - x)(X - \xi)^n \, d\xi = \frac{(X - x)^{n+2}}{(n + 1)(n + 2)}, \tag{4.62}$$

$$\int_0^X (\xi + \beta)(X - \xi)^n \, d\xi = \frac{X^{n+1}[X + (n + 2)\beta]}{(n + 1)(n + 2)}, \tag{4.63}$$

for $n \geq 0$.

We observe that we may commence the above process with the initial estimate

$$T_{-1}(x, X) = 0, \tag{4.64}$$

so that (4.60) and (4.61) yield

$$T_0(x, X) = \left(\frac{X - x}{X + \beta}\right), \qquad t_0(X) = \frac{\alpha}{2} X(X + 2\beta), \tag{4.65}$$

which is simply the pseudo-steady-state approximation. Thus the pseudo-steady-state solution may be considered as a first iteration of the scheme and not simply an arbitrary initial approximation as is commonly supposed. From $(4.65)_1$ and using the integrals (4.62) and (4.63) we find that (4.60) and (4.61) yield

$$T_1(x, X) = \frac{\left\{\alpha\left(\dfrac{X - x}{X + \beta}\right) + \dfrac{1}{2}\left(\dfrac{X - x}{X + \beta}\right)^2 - \dfrac{1}{6}\left(\dfrac{X - x}{X + \beta}\right)^3\right\}}{\left\{\alpha + \dfrac{X(X^2 + 3\beta X + 3\beta^2)}{3(X + \beta)^3}\right\}}, \tag{4.66}$$

$$t_1(X) = \frac{\alpha}{2} X(X + 2\beta) + \frac{X^2(X + 3\beta)}{6(X + \beta)}, \tag{4.67}$$

and we observe $t_1(X)$ is simply the order one corrected motion (see Eqs. (2.41) and (2.77)). Thus, two iterations of (4.64) are seen to produce the known lower and upper bounds $t_0(X)$ and $t_1(X)$, respectively, on the actual motion $t(X)$. A further iteration using $T_1(x, X)$ yields a lengthy expression for $T_2(x, X)$ which we do not give here. However, the corresponding expression for $t_2(X)$ becomes on simplification

$$t_2(X) = \frac{\alpha}{2} X(X + 2\beta) + \frac{X^2\left\{\alpha\left(\dfrac{X + 3\beta}{X + \beta}\right) + X\dfrac{(X^2 + 5\beta X + 5\beta^2)}{5(X + \beta)^3}\right\}}{6\left\{\alpha + \dfrac{X(X^2 + 3\beta X + 3\beta^2)}{3(X + \beta)^3}\right\}}. \tag{4.68}$$

We observe that $T_1(x, X)$ as given by (4.66) is essentially a cubic in $(X - x)$. We may show that T_1 satisfies the inequalities

$$0 \le T_1(x, X) \le T_0(x, X). \tag{4.69}$$

The left-hand inequality follows on inspection, noting $T_0 \le 1$. The right-hand inequality follows since

$$T_1(x, X) - T_0(x, X) = - \frac{T_0\left\{(T_0 - 1)(T_0 - 2) + \dfrac{2X(X^2 + 3\beta X + 3\beta^2)}{(X + \beta)^3}\right\}}{6\left\{\alpha + \dfrac{X(X^2 + 3\beta X + 3\beta^2)}{3(X + \beta)^3}\right\}}$$

$$\le 0. \tag{4.70}$$

Having established (4.69) it follows from (4.61) that at least $t_2(X)$ satisfies known bounds, namely

$$t_0(X) \le t_2(X) \le t_1(X), \tag{4.71}$$

although of course the lower bound is immediately apparent from (4.68).

In order to give some indication of the accuracy of (4.68), numerical values of $t_2(X)/\alpha\beta^2$ are given in Table 4.1 for $\alpha = 2$, 4 and 6 and for the twelve values of X/β used by Pedroso and Domoto (1973a). These values are compared with those obtained from (2.78) using the numerical values of the coefficients $\bar{t}_i(X/\beta)$ ($i = 1, \ldots, 9$) given by Pedroso and Domoto (1973a). For $\alpha = 2$, 4 and 6 the values of $t/\alpha\beta^2$ from (2.78) should be reasonably accurate. It is apparent from Table 4.1 that at least for these values of α, $t_2(X)$ given by (4.68) is a good approximation and appears to be a lower bound to the motion given by Pedroso and Domoto (1973a). We remark that for $\alpha = 1$, (4.68) is not a consistent lower bound to (2.78) but in this case the series (2.78) is much more slowly converging and therefore the values obtained are not as reliable as those for larger α.

4.4.1 Spherical freezing.

For spherical freezing we may deduce from (3.21) and (3.22) the following integral equation for $T^\dagger(r, R)$,

$$T^\dagger(r, R) = \frac{\dfrac{\partial}{\partial R}\left(\displaystyle\int_R^r \xi^2\left(\dfrac{1}{\xi} - \dfrac{1}{r}\right)[\alpha + T^\dagger(\xi, R)]\,d\xi\right)}{\dfrac{\partial}{\partial R}\left(\displaystyle\int_R^1 \xi[1 + (\beta - 1)\xi][\alpha + T^\dagger(\xi, R)]\,d\xi\right)}, \tag{4.72}$$

Table 4.1 Values of $t/\alpha\beta^2$ as given by (4.68) and (2.78) for twelve values of X/β used by Pedroso and Domoto (1973a)

	$t/\alpha\beta^2$ $(\alpha = 2)$		$t/\alpha\beta^2$ $(\alpha = 4)$		$t/\alpha\beta^2$ $(\alpha = 6)$	
X/β	(4.68)	(2.78)	(4.68)	(2.78)	(4.68)	(2.78)
0.2	0.2285	0.2285	0.2244	0.2244	0.2229	0.2229
0.4	0.5105	0.5106	0.4957	0.4957	0.4906	0.4906
0.6	0.8432	0.8434	0.8126	0.8127	0.8020	0.8020
0.8	1.2251	1.2255	1.1743	1.1744	1.1566	1.1567
1.0	1.6555	1.6559	1.5803	1.5804	1.5542	1.5542
1.4	2.6594	2.6600	2.5243	2.5245	2.4774	2.4774
1.8	3.8523	3.8531	3.6432	3.6434	3.5705	3.5706
2.2	5.2328	5.2339	4.9363	4.9365	4.8333	4.8333
2.6	6.8003	6.8016	6.4031	6.4034	6.2652	6.2654
3.0	8.5540	8.5554	8.0434	8.0437	7.8663	7.8664
4.0	13.7520	13.7537	12.9025	12.9027	12.6081	12.6082
5.0	20.1100	20.1123	18.8437	18.8441	18.4053	18.4054

so that from this equation and (3.23) we may formulate the iterative scheme,

$$T_{n+1}(r, R) = \frac{\dfrac{\partial}{\partial R}\left(\displaystyle\int_R^r \xi^2\left(\frac{1}{\xi} - \frac{1}{r}\right)[\alpha + T_n(\xi, R)]\,d\xi\right)}{\dfrac{\partial}{\partial R}\left(\displaystyle\int_R^1 \xi[1 + (\beta - 1)\xi][\alpha + T_n(\xi, R)]\,d\xi\right)}, \qquad (4.73)$$

$$t_{n+1}(R) = \int_R^1 \xi[1 + (\beta - 1)\xi][\alpha + T_n(\xi, R)]\,d\xi, \qquad (4.74)$$

for successive estimates of the temperature and motion, respectively. Again with (4.64) as an initial estimate of the temperature, (4.73) and (4.74) give rise to the pseudo-steady-state temperature and motion for $T_0(r, R)$ and $t_0(R)$ as given by (1.70) and (1.71), respectively. A further application of (4.73) and (4.74) yields, after a calculation similar to the plane problem,

$$T_1(r, R) = \frac{(r - R)\{6\alpha R[1 + (\beta - 1)R]^2 + (r - R)[3 + (\beta - 1)(r + 2R)]\}}{2r\left\{\begin{array}{l}3\alpha R[1 + (\beta - 1)R]^3 + 3\beta(1 - R)[1 + (\beta - 1)R] \\ \qquad\qquad\qquad + (\beta - 1)^2(1 - R)^3\end{array}\right\}} \qquad (4.75)$$

$$t_1(R) = \frac{\alpha}{6}(1 - R)[(1 + 2\beta)(1 + R) + 2(\beta - 1)R^2]$$

$$+ \frac{(1 - R)^2[1 + 2\beta + (\beta - 1)R]}{6[1 + (\beta - 1)R]}, \qquad (4.76)$$

and again $t_1(R)$ is simply the order one corrected motion (see Eqs (2.51) and (2.93)), which is also a known upper bound for the actual motion. A further application of (4.74) using (4.75) gives

$$t_2(R) = \frac{\alpha}{6}(1-R)[(1+2\beta)(1+R)+2(\beta-1)R^2]$$

$$+\frac{(1-R)^2\left\{\begin{array}{l}5\alpha R[1+(\beta-1)R]^2[1+2\beta+(\beta-1)R]\\+5\beta(1-R)[1+(\beta-1)R]+(\beta-1)^2(1-R)^3\end{array}\right\}}{10\left\{\begin{array}{l}3\alpha R[1+(\beta-1)R]^3+3\beta(1-R)[1+(\beta-1)R]\\+(\beta-1)^2(1-R)^3\end{array}\right\}},$$

(4.77)

and this gives rise to the following estimate for the time to complete freezing:

$$t_2(0) = \frac{\alpha}{6}(1+2\beta) + \frac{1}{10}\left(\frac{1+3\beta+\beta^2}{1+\beta+\beta^2}\right). \tag{4.78}$$

In order to confirm $T_1 \leq T_0$ we may write (4.75) as

$$T_1(r, R) = \frac{T_0(r, R)\left\{\begin{array}{l}6\alpha R[1+(\beta-1)R]^2\\+(r-R)[3+(\beta-1)(r+2R)]\end{array}\right\}}{\left\{\begin{array}{l}6\alpha R[1+(\beta-1)R]^2+6\beta(1-R)\\+2(\beta-1)^2(1-R)^3[1+(\beta-1)R]^{-1}\end{array}\right\}}, \tag{4.79}$$

and the required inequality follows provided

$$(r-R)[3+(\beta-1)(r+2R)]$$
$$\leq 6\beta(1-R)+2(\beta-1)^2(1-R)^3[1+(\beta-1)R]^{-1}, \quad (4.80)$$

for $R \leq r \leq 1$. If this latter inequality holds for $r = 1$, it holds for all r such that $R \leq r \leq 1$. Now for $r = 1$, inequality (4.80) simplifies to become

$$2[1+(\beta-1)R]^2 \leq 5\beta[1+(\beta-1)R]+2(\beta-1)^2(1-R)^2, \tag{4.81}$$

and this reduces to

$$\beta[1+2\beta+(\beta-1)R] \geq 0, \tag{4.82}$$

which is certainly true, noting that $\beta \geq 0$. Thus, we again have the important inequalities

$$0 \leq T_1(r, R) \leq T_0(r, R), \tag{4.83}$$

$$t_0(R) \leq t_2(R) \leq t_1(R), \tag{4.84}$$

since (4.84) follows immediately from (4.83) and (4.74) with $n = 1$.

Although $t_2(R)$ satisfies known bounds we observe, however, that this function has some undesirable properties. For example, for β zero we have

$$t_2(R) = \frac{\alpha}{6}(1-R)^2(1+2R) + \frac{(1-R)^2}{10}\left(\frac{1+5\alpha R}{1+3\alpha R}\right), \tag{4.85}$$

so that

$$\frac{dt_2(R)}{dR} = \frac{-(1-R)}{5(1+3\alpha R)^2}[45(\alpha R)^3 + 45(\alpha R)^2 + 14(\alpha R) + 1 - \alpha], \tag{4.86}$$

from which it is apparent that, for $\alpha > 1$, this derivative vanishes for some R in $(0, 1)$. Since the second derivative is negative for this particular R, it follows that $T_2(r, R)$ has a denominator which vanishes for some R in $(0, 1)$ and therefore both $T_2(r, R)$ and $t_3(R)$ are unbounded for some $\alpha > 1$. Clearly, the convergence of these schemes requires further examination. This phenomenon also occurs for β non-zero and Table 4.2 gives numerical values of $t_2(R)$ for $\alpha = 2$ and $\beta = 1$ for various boundary positions. Table 4.2 also gives numerical values of the improved upper and lower bounds (3.70) and (3.84), respectively. The same quantities are shown graphically in Fig. 4.1 for $\alpha = 2$ and β zero. It is apparent from both the table and the figure that the approximate motion is well behaved except for times close to complete freezing, where the motion is clearly

Table 4.2 Values of approximate spherical motion $t_2(R)$ and improved bounds for $\alpha = 2$ and $\beta = 1$ for various boundary positions

Boundary position R	Improved lower bound (3.84)	Approximate motion (4.77)	Improved upper bound (3.70)
0.00	1.080	1.167	1.500
0.05	1.077	1.177	1.449
0.10	1.067	1.174	1.395
0.15	1.051	1.161	1.339
0.20	1.029	1.138	1.280
0.30	0.968	1.067	1.155
0.40	0.886	0.969	1.020
0.50	0.784	0.847	0.875
0.60	0.663	0.707	0.720
0.70	0.524	0.550	0.555
0.80	0.366	0.379	0.380
0.90	0.192	0.195	0.195

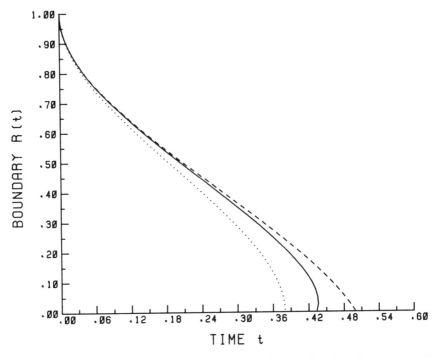

Figure 4.1 Variation of the approximate boundary motion for the sphere for $\alpha = 2$ and β zero compared with improved bounds ($\cdots\cdots$ improved lower bound (3.84), ——— approximate motion (4.77), – – – – improved upper bound (3.70))

not uniquely invertible and the estimate of the boundary velocity becomes infinite. The phenomenon appears to a similar extent for all values of α and β and numerical results seem to indicate that it only ever occurs for small R (approximately $R \leq 0.05$). Finally, we note that the formulae given here for spherical freezing are essentially those presented by Theofanous and Lim (1971) and Shih and Chou (1971). However, because of different notations and terminology the correspondence is not immediately apparent.

4.4.2 Cylindrical freezing. For cylindrical freezing we obtain from (3.25) and (3.26) on eliminating t

$$T^\dagger(r, R) = \frac{\dfrac{\partial}{\partial R}\left(\displaystyle\int_R^r \xi(\log r - \log \xi)[\alpha + T^\dagger(\xi, R)]\,\mathrm{d}\xi\right)}{\dfrac{\partial}{\partial R}\left(\displaystyle\int_R^1 \xi(\beta - \log \xi)[\alpha + T^\dagger(\xi, R)]\,\mathrm{d}\xi\right)}, \tag{4.87}$$

so that in the usual way we may deduce from this equation and (3.27) the following iterative scheme,

$$T_{n+1}(r, R) = \frac{\dfrac{\partial}{\partial R}\left(\displaystyle\int_R^r \xi(\log r - \log \xi)[\alpha + T_n(\xi, R)]\, d\xi\right)}{\dfrac{\partial}{\partial R}\left(\displaystyle\int_R^1 \xi(\beta - \log \xi)[\alpha + T_n(\xi, R)]\, d\xi\right)}, \tag{4.88}$$

$$t_{n+1}(R) = \int_R^1 \xi(\beta - \log \xi)[\alpha + T_n(\xi, R)]\, d\xi. \tag{4.89}$$

As usual with (4.64) as an initial estimate of the temperature, these equations give rise to the pseudo-steady-state estimates (1.73) and (1.74) for $T_0(r, R)$ and $t_0(R)$, respectively. The calculations for further iterations, although straightforward, become rather tedious. The details for $T_1(r, R)$ are simplified somewhat using the equation

$$T_1(r, R) = \frac{\dfrac{\partial}{\partial R}\left(\displaystyle\int_R^r \xi(\log r - \log \xi)\left[\alpha + 1 - \left(\dfrac{\beta - \log \xi}{\beta - \log R}\right)\right] d\xi\right)}{\dfrac{\partial}{\partial R}\left(\displaystyle\int_R^1 \xi(\beta - \log \xi)\left[\alpha + 1 - \left(\dfrac{\beta - \log \xi}{\beta - \log R}\right)\right] d\xi\right)}, \tag{4.90}$$

which becomes

$$T_1(r, R) = \frac{\left\{\begin{array}{l}\alpha R^2(\beta - \log R)^2(\log r - \log R) \\[2mm] \quad + \displaystyle\int_R^r \xi(\log r - \log \xi)(\beta - \log \xi)\, d\xi\end{array}\right\}}{\left\{\alpha R^2(\beta - \log R)^3 + \displaystyle\int_R^1 \xi(\beta - \log \xi)^2\, d\xi\right\}}. \tag{4.91}$$

On evaluating the various integrals we obtain

$$T_1(r, R) = \frac{\left\{\begin{array}{l}R^2[4\alpha(\beta - \log R)^2 - 2(\beta - \log R) - 1][\log r - \log R] \\[2mm] \quad + (1 + \beta)(r^2 - R^2) - (r^2 \log r - R^2 \log R)\end{array}\right\}}{\left\{\begin{array}{l}R^2[4\alpha(\beta - \log R)^3 + 2(1 + 2\beta)\log R - 2(\log R)^2] \\[2mm] \quad + (1 + 2\beta + 2\beta^2)(1 - R^2)\end{array}\right\}}. \tag{4.92}$$

We note that (4.91), in addition to being a useful equation in the derivation of (4.92), is important for two other reasons. Firstly, it is apparent from (4.91) that both the numerator and denominator are greater than or equal to zero and therefore $T_1 \geq 0$. Secondly, we may use

(4.91) to show that $T_1 \le T_0$, since we have

$$T_1(r, R) - T_0(r, R)$$

$$= -\frac{\left\{\begin{array}{l} (\beta - \log r) \int_R^r \xi(\beta - \log \xi)(\log \xi - \log R) \, d\xi \\ \quad + (\log r - \log R) \int_r^1 \xi(\beta - \log \xi)^2 \, d\xi \end{array}\right\}}{(\beta - \log R)\left\{ \alpha R^2(\beta - \log R)^3 + \int_R^1 \xi(\beta - \log \xi)^2 \, d\xi \right\}},$$

$$\le 0, \tag{4.93}$$

and therefore the required result follows. Clearly, these observations are not quite so apparent from the explicit expression (4.92). Thus, again, the inequalities (4.83) and (4.84) hold where $t_1(R)$ is the order one corrected motion (see Eqs (2.61) and (2.107)).

From (4.92) and (4.89) with $n = 1$ we obtain after a long but straight-forward calculation

$$t_2(R) = \frac{\alpha}{4} [2R^2 \log R + (1 + 2\beta)(1 - R^2)]$$

$$+ \frac{\left\{\begin{array}{l} (3 + 12\beta + 8\beta^2)(1 - R^4) + 4R^2[(3 + 4\beta)R^2 + 8\beta(1 + \beta)] \log R \\ - 32\alpha R^2(\beta - \log R)^2[(1 + \beta)(1 - R^2) + (R^2 + 1 + 2\beta) \log R] \\ - 8R^2[R^2 + 2(1 + 2\beta)](\log R)^2 \end{array}\right\}}{\left\{\begin{array}{l} R^2[4\alpha(\beta - \log R)^3 + 2(1 + 2\beta) \log R - 2(\log R)^2] \\ + (1 + 2\beta + 2\beta^2)(1 - R^2) \end{array}\right\}}.$$

$$\tag{4.94}$$

This gives rise to the following estimate for the time to complete freezing,

$$t_2(0) = \frac{\alpha}{4}(1 + 2\beta) + \frac{1}{32}\left(\frac{3 + 12\beta + 8\beta^2}{1 + 2\beta + 2\beta^2}\right), \tag{4.95}$$

which is in agreement with Shih and Tsay (1971).

Numerical results indicate that the approximate motion (4.94) has the same undesirable feature as (4.77) for times close to complete freezing, except that for the cylindrical problem this phenomenon appears to occur sooner (approximately $R \le 0.10$). This is apparent from Table 4.3, which gives numerical values of $t_2(R)$ for $\alpha = 2$ and $\beta = 2$ for various boundary positions, and from Fig. 4.2, which shows the behaviour graphically for $\alpha = 1$ and $\beta = 2$.

4.5 Some general comments

Firstly, it is perhaps worth noting that, in terms of the general equation (1.34) and the function $K_\lambda(x, y)$ defined by (3.95), we may obtain from

Table 4.3   Values of approximate cylindrical motion $t_2(R)$ and improved bounds for $\alpha = 2$ and $\beta = 2$ for various boundary positions

Boundary position R	Improved lower bound (3.91)	Approximate motion (4.94)	Improved upper bound (3.72)
0.00	2.610	2.642	3.750
0.05	2.593	2.697	3.087
0.10	2.553	2.703	2.950
0.15	2.495	2.660	2.824
0.20	2.422	2.585	2.698
0.30	2.236	2.375	2.432
0.40	2.006	2.114	2.143
0.50	1.740	1.816	1.831
0.60	1.441	1.491	1.498
0.70	1.115	1.143	1.146
0.80	0.764	0.777	0.777
0.90	0.391	0.395	0.395

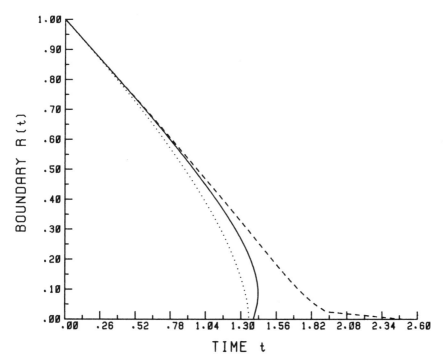

Figure 4.2   Variation of the approximate boundary motion for the cylinder for $\alpha = 1$ and $\beta = 2$ compared with improved bounds ($\cdots\cdots$ improved lower bound (3.91), ——— approximate motion (4.94), $-----$ improved upper bound (3.72))

(3.101) and (3.103) the following general formulation for the integral equation underlying the iteration scheme,

$$T^\dagger(r, R) = \frac{\dfrac{\partial}{\partial R}\left(\displaystyle\int_R^r \xi^\lambda K_\lambda(r, \xi)[\alpha + T^\dagger(\xi, R)]\,d\xi\right)}{\dfrac{\partial}{\partial R}\left(\displaystyle\int_R^1 \xi^\lambda[\beta + K_\lambda(1, \xi)][\alpha + T^\dagger(\xi, R)]\,d\xi\right)}. \tag{4.96}$$

In particular, with $T_0(r, R)$ as the pseudo-steady-state estimate given by (3.96) we may confirm that $T_1(r, R)$ becomes

$$T_1(r, R) = \frac{\left\{\begin{array}{c}\alpha R^{2\lambda}K_\lambda(r, R)[\beta + K_\lambda(1, R)]^2 \\[2mm] + \displaystyle\int_R^r \xi^\lambda K_\lambda(r, \xi)[\beta + K_\lambda(1, \xi)]\,d\xi\end{array}\right\}}{\left\{\alpha R^{2\lambda}[\beta + K_\lambda(1, R)]^3 + \displaystyle\int_R^1 \xi^\lambda[\beta + K_\lambda(1, \xi)]^2\,d\xi\right\}}, \tag{4.97}$$

and it is apparent from this formulation that T_1 satisfies the inequalities (4.83). Clearly, $T_1 \geq 0$ and further since

$$T_1(r, R) - T_0(r, R)$$

$$= -\frac{\left\{\begin{array}{c}[\beta + K_\lambda(1, r)]\displaystyle\int_R^r \xi^\lambda[\beta + K_\lambda(1, \xi)]K_\lambda(\xi, R)\,d\xi \\[2mm] + K_\lambda(r, R)\displaystyle\int_r^1 \xi^\lambda[\beta + K_\lambda(1, \xi)]^2\,d\xi\end{array}\right\}}{[\beta + K_\lambda(1, R)]\left\{\alpha R^{2\lambda}[\beta + K_\lambda(1, R)]^3 + \displaystyle\int_R^1 \xi^\lambda[\beta + K_\lambda(1, \xi)]^2\,d\xi\right\}},$$

$$\leq 0, \tag{4.98}$$

and it follows that $T_1 \leq T_0$. As noted previously for the cylindrical geometry, it is interesting that these important inequalities are more apparent from the general structure than explicit representations. On the other hand, it appears to be difficult to make any further progress in general terms.

Secondly, the unusual nature of the integral equation (4.96) is essentially due to the necessity of simultaneously estimating both the temperature and boundary position. In view of the integral formula (3.102) for the motion of the boundary, an alternative strategy might be to attempt to obtain separately the 'best possible' approximate solution of the heat equation, subject only to the surface and boundary conditions (2.5), and then utilize this estimate with (3.102) to obtain the boundary motion.

Further, in view of the simple nature of the Green's function (3.99) we might consider the alternative iterative scheme for $\bar{T}_n(r, R)$

$$\frac{\partial^2 \bar{T}_n}{\partial r^2} + \frac{\lambda}{r} \frac{\partial \bar{T}_n}{\partial r} = \frac{\partial \bar{T}_{n-1}}{\partial t}, \qquad R(t) < r < 1, \tag{4.99}$$

subject to both conditions (2.5). In this case we have

$$\bar{T}_n(r, R) = \bar{T}_{n-1}(r, R)$$
$$+ \int_{R(t)}^1 G(r, \xi, t) \xi^\lambda \left\{ \frac{\partial \bar{T}_{n-1}}{\partial t} (\xi, R(t)) - \frac{\partial \bar{T}_{n-2}}{\partial t} (\xi, R(t)) \right\} d\xi, \tag{4.100}$$

where $G(r, \xi, t)$ is defined by (3.99). In particular, with $\bar{T}_{-1}(r, R) \equiv 0$ and $\bar{T}_0(r, R)$ as the pseudo-steady-state estimate (3.96) we have

$$\bar{T}_1(r, R) = \bar{T}_0(r, R) - \dot{R}(t) \int_R^1 \frac{G(r, \xi, t) \xi^\lambda [\beta + K_\lambda(1, \xi)] \, d\xi}{R^\lambda [\beta + K_\lambda(1, R)]^2}, \tag{4.101}$$

and since both $G(r, \xi, t)$ and $\dot{R}(t)$ are both less than or equal to zero it follows immediately from (4.101) that $\bar{T}_1 \le \bar{T}_0$ but not that $\bar{T} \ge 0$. We might now substitute (4.101) into (3.102) to obtain an ordinary differential equation for $R(t)$, which in principle can be solved subject to $R(0) = 1$ to give an estimate of the motion of the boundary. However, since the resulting ordinary differential equation is by no means straightforward to solve, we are again forced into the situation of estimating $\dot{R}(t)$ for use in (4.101). It is of interest to note that if we use the pseudo-steady-state estimate of $\dot{R}(t)$, namely

$$\dot{R}(t) \simeq -\{\alpha R^\lambda [\beta + K_\lambda(1, R)]\}^{-1}, \tag{4.102}$$

then (4.101) yields the temperature profile correct up to order α^{-1}, and substitution of this estimate into (3.102) yields the correct motion up to order α^{-2} (see Chapter 2). On the other hand, if we use the first order corrected estimate of $\dot{R}(t)$, namely

$$\dot{R}(t) \simeq -\left\{ \alpha R^\lambda [\beta + K_\lambda(1, R)] + \int_R^1 \frac{\xi^\lambda [\beta + K_\lambda(1, \xi)]^2}{R^\lambda [\beta + K_\lambda(1, R)]^2} \, d\xi \right\}^{-1}, \tag{4.103}$$

then (4.101) gives rise to precisely the approximation $T_1(r, R)$ discussed in the previous section (see in particular the expression given in (4.98)). Thus, in terms of iterations of the pseudo-steady-state estimate, the standard approach yielding (4.97) might be the simplest and best that can be achieved.

4.6 Additional symbols used

$c_n(z, z_0)$, $e_n(z, z_0)$	Langford's cylindrical functions, (4.2) and (4.3)
$f(t)$	arbitrary function of time such that (4.31) holds
p	Laplace transform variable, (4.32)
$R^{-1}(r)$	inverse of $r = R(t)$ for sphere and cylinder, (4.44)
$t_n(X)$, $t_n(R)$	successive estimates for boundary motion, (4.61) and (4.74)
$T_n(x, X)$, $T_n(r, R)$	successive estimates for temperature, (4.60) and (4.73)
$\hat{T}(x, p)$, $\hat{T}(r, p)$	modified Laplace transforms, (4.40) and (4.44)
$\bar{T}_n(r, R)$	successive estimates for non-standard iterative scheme, (4.99)
$X^{-1}(x)$	inverse of $x = X(t)$, (4.39)
z, z_0	denotes $r^2/4$ and $r_0^2/4$, (4.1)
$Z(t)$	denotes $R(t)^2/4$, (4.23)

Greek symbols

$\phi(x, y)$, $\psi(x, y)$	functions defined by (4.54) and (4.55)
η, ξ, ω	integration variables

Convention

Dot denotes differentiation with respect to time t.

5

Polynomial approximations

5.1 Introduction

In this chapter we examine various quadratic and cubic approximations to the temperature profiles. For example, since the pseudo-steady-state estimate (1.66) of the problem (2.1)–(2.3) is linear in $(x - X(t))$ it is natural to inquire whether a better approximation may be achieved using higher-order polynomials in the variable $(x - X(t))$. Thus, for the problem (2.1)–(2.3) we might seek to approximate the temperature profile by an expression of the form

$$T(x, t) = \sum_{n=1}^{N} A_n(t)(x - X(t))^n, \qquad (5.1)$$

where $A_n(t)$ are as yet unknown functions of time and N is a positive integer. If we attempt to determine the $A_n(t)$ by substituting (5.1) into (2.1), then a system of ordinary differential equations is obtained for these functions, and the resulting series (5.1) with N infinite is a rearrangement of the series given by (4.10). If, however, N is finite, (2.1) cannot be satisfied in an exact sense, and the functions $A_n(t)$ are determined by satisfying surface and boundary conditions and requiring that (2.1) and its time derivatives are satisfied at the end points of the interval. In this chapter we consider the main quadratic and cubic approximations to the three Stefan problems for planar, spherical and cylindrical freezing. We show that using the pseudo-steady-state solution as a basis for the polynomial approximation is not always effective, and for the cylindrical problem we propose an alternative temperature profile which is not a polynomial. The form of this alternative profile is motivated by the formal series solution (4.29) and provides a non-trivial application of (4.29).

In the following section, in order to introduce the various quadratic approximations in the simplest way, we consider the problem of planar solidification (2.1)–(2.3) with β zero, so that the possible approximations to the motion of the boundary can be contrasted with the exact solution $X(t) = (2\gamma t)^{1/2}$. For a quadratic temperature profile three possibilities for the boundary motion present themselves. These correspond to the utilization of either the Stefan condition, the exact integral formulae of Chapter 3 or the satisfaction of the heat-conduction equation in an average or integral sense. Utilization of the Stefan condition is sometimes known in the literature as Megerlin's method (see Solomon (1978) and subsequent related papers by the same author and also Goodling and Khader (1975)). However, despite its relative simplicity Megerlin's procedure is not widely used in the literature. On the other hand, satisfaction of the heat conduction equation in an integral sense (that is, the heat-balance method) has been employed and extended to numerous problems. This procedure in the context of phase-change problems is due primarily to Goodman (1958, 1961) (see also the review article by Goodman (1964) and Goodman and Shea (1960)). Despite the popularity of the heat-balance method we see in Section 5.3 that this procedure for the planar problem with surface cooling is far more complicated than either Megerlin's method or the integral formulae approach and, moreover, the latter methods appear to be more accurate. Accordingly, for spherical and cylindrical freezing given in Sections 5.4 and 5.5, respectively, we only give the heat-balance approximation for the case of no surface cooling. Further, the heat-balance method for the quadratic logarithmic profile (5.63) predicts an infinite solidification time for the cylindrical problem (see Eq. (5.78)). We note here that for cylindrical problems Sparrow (1960) reports the inadequacy of polynomial profiles while Lardner and Pohle (1961) propose a single-term logarithmic profile. In Section 5.6 we illustrate the use of cubic temperature profiles with reference to the three freezing problems but only for the case of no surface cooling (that is β zero). In Section 5.7 we consider the alternative approximating temperature profile for the cylindrical problem which is motivated by the formal exact series solution (4.29).

It will become apparent to the reader that the methods described in this chapter are arbitrary to a considerable extent. Firstly, with respect to the choice of the approximating temperature profile and, secondly, with respect to the conditions employed for the determination of the coefficients involved in the profile. For example, for the spherical problem two possible approximating profiles are (5.30) and (5.58), which of course give varying results. With regard to the second matter, the basic idea is that we wish to avoid time differentiations of the approximating temperature profile. This may be achieved by differentiating, with respect to time, the surface and boundary conditions (as many times as is necessary) and then using the partial differential equation to translate this information into conditions on spatial derivatives at the end points. Even with this in mind there is still a good deal of freedom as to which are the important conditions to be satisfied at the end points. Goodman (1961) states the following fundamental rule:

In selecting a derived constraint at one end of the interval in order to determine an additional constant in the profile, the accuracy will be improved only if preference is given to the one which involves the lower order derivative. If the highest order derivative involved in both possibilities is the same, then the choice is arbitrary.

Goodman notes that this rule is probably correct since, in approximating the temperature profile, the priorities at the end points are first to obtain agreement with the temperature values themselves, then the first derivatives, then the second derivatives and so on. Goodman discusses this rule in the context of the complete mathematical breakdown of a heat-balance solution using a quartic approximation and we refer the reader to the paper for the details. Due to the extensive use of the heat-balance method in the literature, for completeness, we close this section with a brief review of some of the extensions and applications of the method and further references can be found in the articles cited.

The review article by Goodman (1964) along with the papers previously cited by the same author constitute by far the best introductory account of the method with applications to a variety of interesting melting problems. In addition, the review article summarizes related integral methods and techniques of improving the accuracy. Goodman and Shea (1960) and Yuen (1980) apply extensions of the heat-balance method to genuine two-phase melting problems. The first paper reduces the problem to a coupled system of non-linear ordinary differential equations which need to be solved either numerically or by some approximate asymptotic method. The second paper avoids these complexities by expressing the energy-balance condition at the front in two ways, from which a closed form approximation may be obtained. Duck and Riley (1977) and Riley and Duck (1977) extend the approach of Poots (1962a) to multi-dimensional problems including ellipses, ellipsoids, rectangular prisms and cuboids. Other single-component applications or refinements of the heat-balance method are examined by Bell (1978, 1979), Gutfinger and Chen (1969) and Hameed and Lebedeff (1975), while various multi-component applications are considered by Essoh and Klinzing (1980), Grange *et al.* (1976) and Gupta (1974). Finally, we note that Poots (1962b) describes integral methods for solidification problems as adapted from similar techniques employed in fluid mechanics.

5.2 Quadratic approximations

In order to give a simple introduction to approximation by quadratic temperature profiles we consider planar freezing with β zero. We seek to approximate the solution of (2.1)–(2.3) by an expression of the form

$$T(x, t) = A_1(t)(x - X(t)) + A_2(t)(x - X(t))^2, \tag{5.2}$$

where $A_1(t)$ and $A_2(t)$ are determined from known information at the end points $x = 0$ and $x = X(t)$ in the following manner. On differentiating $T(X(t), t) = 0$ we have

$$\frac{\partial T}{\partial t}(X(t), t) + \frac{\partial T}{\partial x}(X(t), t)\frac{dX}{dt} = 0, \tag{5.3}$$

and from this equation and (2.3)$_1$ we find that Eq. (2.1) at $x = X(t)$ becomes

$$\left(\frac{\partial T}{\partial x}(X(t), t)\right)^2 = \alpha\frac{\partial^2 T}{\partial x^2}(X(t), t), \tag{5.4}$$

which, incidentally, is the essential non-linearity in the problem. From (5.2) and (5.4) we readily obtain the relation

$$A_2(t) = \frac{A_1(t)^2}{2\alpha}. \tag{5.5}$$

From (5.2), (5.5) and the surface condition $T(0, t) = 1$ we obtain the following quadratic equation for the determination of $A_1(t)$, thus

$$A_1(t)^2 X(t)^2 - 2\alpha A_1(t)X(t) - 2\alpha = 0, \tag{5.6}$$

where the two roots correspond to the situations of freezing in the negative or positive half-planes. This can be seen from Eq. (5.8) from which it is apparent that $A_1(t)$ must be negative. Alternatively, since $x < X(t)$ it is clear from (5.2) that taking the negative root of (5.6) ensures $T \geq 0$. Thus we have

$$A_1(t) = \frac{\alpha - (\alpha^2 + 2\alpha)^{1/2}}{X(t)}, \qquad A_2(t) = \frac{1 + \alpha - (\alpha^2 + 2\alpha)^{1/2}}{X(t)^2}, \tag{5.7}$$

and it follows from these equations and (5.2) that in this case we have simply proposed a temperature profile which is quadratic in the pseudo-steady-state temperature estimate (1.66). We find in subsequent sections that polynomial temperature profiles utilizing the pseudo-steady-state solution are not always effective and frequently exhibit the same peculiarities which are characteristic of the pseudo-steady-state solution. For example, the cylindrical profile (5.63) and the alternative spherical profile

(5.58) both predict the physically unrealistic result $T(r, t_c) = 1$, which is a characteristic of the pseudo-steady-state estimates (1.73) and (1.70), respectively, even though for β non-zero both profiles are not simply quadratic in the corresponding pseudo-steady-state temperature.

Up to Eq. (5.7), the three quadratic approximations considered here all agree. However, at this stage there is no unique procedure to determine the boundary motion $X(t)$ and three possibilities arise. For the problem under consideration the three possibilities all give rise to boundary motions of the form $X(t) = (2\gamma t)^{1/2}$ but with various estimates of the constant γ.

5.2.1   Megerlin's method.

In this case we obtain $X(t)$ from the Stefan condition $(2.3)_1$, which from (5.2) becomes

$$A_1(t) = -\alpha \frac{dX}{dt}. \tag{5.8}$$

Thus, from (5.7) and the initial condition $(2.3)_2$ we readily obtain $X(t) = (2\gamma t)^{1/2}$ with the following estimate for γ^{-1},

$$\gamma^{-1} \simeq \frac{\alpha}{2}\left[1 + \left(1 + \frac{2}{\alpha}\right)^{1/2}\right]. \tag{5.9}$$

5.2.2   Integral formula method.

If instead of using the Stefan condition we employ the integral formula (3.6) with β zero, then from (5.2) and (5.5) we obtain

$$t = \frac{X(t)^2}{2}\left[\alpha - \frac{A_1(t)X(t)}{3} + \frac{A_1(t)^2 X(t)^2}{12\alpha}\right], \tag{5.10}$$

which using (5.7) gives rise to the following estimate for γ^{-1}:

$$\gamma^{-1} \simeq \frac{\alpha}{6}\left[5 + \left(1 + \frac{2}{\alpha}\right)^{1/2}\right] + \frac{1}{6}. \tag{5.11}$$

5.2.3   Heat-balance method.

On integrating (2.1) from the surface $x = 0$ to the moving boundary $x = X(t)$ we obtain (see Eq. (3.1))

$$\frac{d}{dt}\left(\int_0^{X(t)} T(\xi, t)\,d\xi + \alpha X(t)\right) + \frac{\partial T}{\partial x}(0, t) = 0. \tag{5.12}$$

On using (5.2) and (5.5) this equation becomes

$$\frac{d}{dt}\left[\alpha X(t) - \frac{A_1(t)X(t)^2}{2} + \frac{A_1(t)^2 X(t)^3}{6\alpha}\right] + A_1(t) - \frac{A_1(t)^2 X(t)}{\alpha} = 0, \tag{5.13}$$

which on using (5.7) can be integrated to give the following estimate for γ^{-1}:

$$\gamma^{-1} \simeq \frac{1}{2}\left(\alpha + \frac{1}{3}\right)\left[1 + \left(\frac{\alpha}{\alpha+2}\right)^{1/2}\right] + \frac{1}{6}\left(\frac{\alpha}{\alpha+2}\right)^{1/2}. \tag{5.14}$$

Table 5.1 lists the various estimates for γ^{-1}. Also included are the estimates for the pseudo-steady-state, the three-term large α expansion, the integral iteration approximation and the estimates (5.86) and (5.88) arising from the cubic temperature profile (5.79). Table 5.2 gives numerical values of the various estimates of γ^{-1} compared with the exact value which is obtained as a root of the transcendental equation (1.52). It is apparent from this table that the integral iteration estimate is uniformly the best approximation, while the integral formula estimate gives the next best estimate and is by far the best result out of the three quadratic approximations. For $\alpha \geq 1$ the proximity with respect to the exact value is as indicated in the table. These results tend to suggest that the additional work involved in either the heat-balance method with a quadratic profile

Table 5.1   Estimates of γ^{-1} for various approximations

Name of approximation	Estimate of γ^{-1}	Equation
Pseudo-steady-state	α	(1.67)
Large α expansion	$\alpha + \frac{1}{3} - \dfrac{2}{45\alpha}$	(2.75)
Integral iteration	$\alpha + \frac{1}{3} - \dfrac{2}{45(\alpha + \frac{1}{3})}$	(4.68)
Megerlin (quadratic)	$\dfrac{\alpha}{2}\left[1 + \left(1 + \dfrac{2}{\alpha}\right)^{1/2}\right]$	(5.9)
Integral formula (quadratic)	$\dfrac{\alpha}{6}\left[5 + \left(1 + \dfrac{2}{\alpha}\right)^{1/2}\right] + \dfrac{1}{6}$	(5.11)
Heat balance (quadratic)	$\frac{1}{2}(\alpha + \frac{1}{3})\left[1 + \left(\dfrac{\alpha}{\alpha+2}\right)^{1/2}\right] + \dfrac{1}{6}\left(\dfrac{\alpha}{\alpha+2}\right)^{1/2}$	(5.14)
Megerlin (cubic)	$\dfrac{\alpha}{2}\left[1 + \left(1 + \dfrac{4}{3\alpha}\right)^{1/2}\right]$	(5.86)
Integral formula (cubic)	$\dfrac{\alpha}{5}\left[4 + \left(1 + \dfrac{4}{3\alpha}\right)^{1/2}\right] + \dfrac{1}{5}$	(5.88)

Table 5.2 Numerical values of the various estimates for γ^{-1} compared with the exact value

α	Heat balance (quadratic) (5.14)	Megerlin (cubic) (5.86)	Integral formula (quadratic) (5.11)	Large α (three terms) (2.75)	Integral iteration (4.68)	Exact γ^{-1} (1.52)	Integral formula (cubic) (5.88)	Megerlin (quadratic) (5.9)
0.1	0.30	0.24	0.33	Negative	0.33	0.32	0.36	0.28
0.2	0.40	0.38	0.44	0.31	0.45	0.45	0.47	0.43
0.5	0.68	0.73	0.77	0.74	0.78	0.78	0.80	0.81
1.0	1.15	1.26	1.29	1.29	1.30	1.30	1.31	1.37
2.0	2.11	2.29	2.31	2.31	2.31	2.32	2.32	2.41
5.0	5.06	5.31	5.32	5.33	5.33	5.33	5.33	5.46
10.0	10.04	10.32	10.33	10.33	10.33	10.33	10.33	10.48
20.0	20.02	20.33	20.33	20.33	20.33	20.33	20.33	20.49

or going to a cubic temperature profile is not really justified. Further, they tend to indicate that methods based on an integral formulation of the problem are superior. In the following section we extend the calculations given here to the case of β non-zero. The analysis becomes considerably complicated, especially for the heat-balance method.

5.3 Planar solidification with surface cooling

For the problem of planar solidification with β non-zero, Eqs. (5.3) and (5.4) hold and assuming the quadratic profile (5.2) we still have the relation (5.5). However, from the surface condition $(2.2)_1$ we now have the quadratic equation

$$A_1(t)^2 X(X + 2\beta) - 2\alpha A_1(t)(X + \beta) - 2\alpha = 0, \tag{5.15}$$

where for brevity here and subsequently we omit the explicit time dependence in the boundary motion $X(t)$. The appropriate solution for $A_1(t)$ ensuring $T \geq 0$ is

$$A_1(t) = \frac{\alpha(X + \beta) - [\alpha^2(X + \beta)^2 + 2\alpha X(X + 2\beta)]^{1/2}}{X(X + 2\beta)}, \tag{5.16}$$

and the details for the three approximations to the boundary now follow.

5.3.1 Megerlin's method.   From the Stefan condition $(2.3)_1$ and (5.2) we again obtain (5.8), which on using (5.16) simplifies to give

$$2 \, dt = \{\alpha(X + \beta) + [\alpha^2(X + \beta)^2 + 2\alpha X(X + 2\beta)]^{1/2}\} \, dX, \tag{5.17}$$

and on integrating this equation and using the initial condition $X(0) = 0$ we have

$$\begin{aligned}
t = \frac{1}{4} \Big\{ &\alpha X(X + 2\beta) + (X + \beta)[\alpha^2(X + \beta)^2 + 2\alpha X(X + 2\beta)]^{1/2} - \alpha\beta^2 \\
&- 2\beta^2 \left(\frac{\alpha}{\alpha + 2}\right)^{1/2} \\
&\times \log\left[\frac{(X + \beta)(\alpha^2 + 2\alpha)^{1/2} + [\alpha^2(X + \beta)^2 + 2\alpha X(X + 2\beta)]^{1/2}}{\alpha\beta + \beta(\alpha^2 + 2\alpha)^{1/2}}\right] \Big\}.
\end{aligned} \tag{5.18}$$

We observe that the pseudo-steady-state approximation (1.67) emerges for large α and that (5.9) arises from (5.18) in the limit β tending to zero.

5.3.2 *Integral formula method.* From (3.6), (5.2) and (5.5) we obtain

$$t = \frac{\alpha}{2}X(X+2\beta) - \frac{A_1(t)X^2}{6}(X+3\beta) + \frac{A_1(t)^2X^3}{24\alpha}(X+4\beta), \qquad (5.19)$$

and substituting (5.16) into this expression gives

$$t = \frac{X}{12(X+2\beta)^2}\Big\{\alpha(5X^3 + 30\beta X^2 + 59\beta^2 X + 40\beta^3) + X(X+2\beta)(X+4\beta)$$

$$+ (X^2 + 5\beta X + 8\beta^2)[\alpha^2(X+\beta)^2 + 2\alpha X(X+2\beta)]^{1/2}\Big\}. \qquad (5.20)$$

Again this gives $\alpha X(X+2\beta)/2$ for large α and is consistent with (5.11).

5.3.3 *Heat-balance method.* If we introduce $C(t)$ such that

$$C(t) = A_1(t)X(t), \qquad (5.21)$$

then it is not difficult to show that (5.13), which still holds for β non-zero, becomes

$$\frac{d}{dt}\left[\frac{X}{2}\left(\frac{C^2}{3\alpha} - C\right) + \alpha X\right] = \frac{1}{X}\left(\frac{C^2}{\alpha} - C\right), \qquad (5.22)$$

and

$$C(t) = \frac{\alpha(X+\beta) - [\alpha^2(X+\beta)^2 + 2\alpha X(X+2\beta)]^{1/2}}{(X+2\beta)}. \qquad (5.23)$$

Further, if we introduce $D(t)$ by

$$D(t) = C(t) - \alpha, \qquad (5.24)$$

then we have

$$X(t) = \frac{-2\beta D(D+\alpha)}{(D^2 - b^2)}, \qquad (5.25)$$

where $b = (\alpha^2 + 2\alpha)^{1/2}$ and (5.22) becomes

$$\frac{d}{dt}\left[\frac{D(D+\alpha)(D^2 - \alpha D + 4\alpha^2)}{3(D^2 - b^2)}\right] = \frac{D^2 - b^2}{2\beta^2}. \qquad (5.26)$$

On differentiating and rearranging this equation the problem of integrating (5.13) reduces to the problem of integrating the following:

$$\frac{3}{4\beta^2}\,dt = \left\{\frac{D}{(D^2 - b^2)} - \frac{2\alpha^3}{(D^2 - b^2)^2} - \frac{2\alpha^2(\alpha+2)(2\alpha+1)D}{(D^2 - b^2)^3}\right.$$

$$\left. - \frac{4\alpha^4(\alpha+2)}{(D^2 - b^2)^4}\right\}dD. \qquad (5.27)$$

On integrating this equation we obtain

$$
\frac{3t}{2\beta^2} = \left\{ \log|D^2 - b^2| - \frac{\alpha^2 D}{(\alpha + 2)(D^2 - b^2)} \right.
$$
$$
\left. + \frac{[\alpha^2(\alpha + 2)(2\alpha + 1) + 2\alpha^3 D]}{(D^2 - b^2)^2} + \frac{1}{2}\left(\frac{\alpha}{\alpha + 2}\right)^{3/2} \log\left|\frac{D + b}{D - b}\right| \right\}
$$
$$
+ \text{ constant of integration.} \tag{5.28}
$$

After much simplification and choosing the constant of integration appropriately we finally obtain

$$
t = \frac{1}{12(\alpha + 2)} \left\{ [3\alpha(X + \beta) + 2(X + 2\beta)][\alpha^2(X + \beta)^2 + 2\alpha X(X + 2\beta)]^{1/2} \right.
$$

$$
+ (\alpha + 2)(3\alpha + 1)X^2 + 2(3\alpha^2 + 12\alpha + 4)\beta X - \alpha(3\alpha + 4)\beta^2
$$

$$
- 8(\alpha + 2)\beta^2 \log\left[\frac{\begin{array}{c}\alpha(X + \beta) + 2(X + 2\beta)\\ + [\alpha^2(X + \beta)^2 + 2\alpha X(X + 2\beta)]^{1/2}\end{array}}{2(\alpha + 2)\beta}\right]
$$

$$
\left. - \frac{4\alpha_2^3\beta^2}{(\alpha + 2)^{1/2}} \log\left[\frac{\begin{array}{c}(X + \beta)(\alpha^2 + 2\alpha)^{1/2}\\ + [\alpha^2(X + \beta)^2 + 2\alpha X(X + 2\beta)]^{1/2}\end{array}}{\alpha\beta + \beta(\alpha^2 + 2\alpha)^{1/2}}\right] \right\}.
$$
$$
\tag{5.29}
$$

This equation agrees with that given by Goodman (1958) except that in the second logarithmic term Goodman has $[(X + \beta)\alpha(\alpha + 2)]^{1/2}$ instead of $(X + \beta)(\alpha^2 + 2\alpha)^{1/2}$. Clearly, for β non-zero the heat-balance method is considerably more complicated and will not be attempted for spherical and cylindrical freezing. Again (5.29) gives the pseudo-steady-state result for large α and agrees with (5.14) in the limit as β tends to zero.

Table 5.3 gives numerical values of $t/\alpha\beta^2$ for the three approximations and for twelve values of X/β used by Pedroso and Domoto (1973a). Again it is apparent that the integral formula provides the best estimate of the three quadratic approximations.

5.4   Freezing a liquid in a spherical container

For polynomial approximations to spherically symmetric problems there are clearly at least two possible approaches. Either we can make the standard transformation (5.45) and adopt a polynomial profile for $v(r, t)$

Table 5.3 Numerical values of $t/\alpha\beta^2$ for the three planar quadratic approximations compared with the values given by Pedroso and Domoto (1973a) for twelve values of X/β and $\alpha = 2$

X/β	Heat balance (5.29)	Integral formula (5.20)	Pedroso and Domoto (2.78)	Megerlin (5.18)
0.2	0.23	0.23	0.23	0.23
0.4	0.51	0.51	0.51	0.51
0.6	0.84	0.84	0.84	0.85
0.8	1.22	1.22	1.23	1.24
1.0	1.65	1.65	1.66	1.68
1.4	2.63	2.65	2.66	2.71
1.8	3.80	3.84	3.85	3.94
2.2	5.14	5.22	5.24	5.36
2.6	6.65	6.78	6.80	6.98
3.0	8.34	8.53	8.56	8.80
4.0	13.31	13.70	13.75	14.19
5.0	19.35	20.04	20.11	20.80

based on the planar pseudo-steady-state solution or we might express $T(r, t)$ directly as a polynomial in the spherical pseudo-steady-state solution. Thus, for the spherical freezing problem (2.4)–(2.6) we might employ either

$$T(r, t) = \frac{1}{r}\left\{A_1(t)(r - R) + A_2(t)(r - R)^2\right\},$$ (5.30)

where again for brevity we have omitted the explicit time dependence of the boundary $R(t)$, or the alternative quadratic temperature profile (5.58), details of which are noted at the end of this section. In this section we have adopted (5.30) in preference to (5.58), firstly, on the basis that (5.30) predicts the physically more realistic temperature profile (5.36) at complete freezing (in contrast to (5.58) which predicts $T(r, t_c) = 1$) and, secondly, on the basis that $T(r, t_c) = 1$ in conjunction with the integral formula (3.23) yields no more that the upper bound in (3.62) as the estimate of the time t_c.

On differentiating (2.5)$_2$ and using (2.4) we have

$$\left(\frac{\partial T}{\partial r}(R, t)\right)^2 = \alpha\left(\frac{\partial^2 T}{\partial r^2}(R, t) + \frac{2}{R}\frac{\partial T}{\partial r}(R, t)\right),$$ (5.31)

and from (5.30) and (5.31) we readily obtain

$$A_2(t) = \frac{A_1(t)^2}{2\alpha R}.$$ (5.32)

From the surface condition (2.5)$_1$ and the above equations we may deduce the quadratic equation for $A_1(t)$,

$$A_1(t)^2(1-R)[1+\beta+(\beta-1)R]+2\alpha RA_1(t)[1+(\beta-1)R]-2\alpha R=0,$$
(5.33)

so that the appropriate solution ensuring $T \geq 0$ is

$$A_1(t) = \frac{\left\{\begin{array}{c}[\alpha^2 R^2[1+(\beta-1)R]^2+2\alpha R(1-R)[1+\beta+(\beta-1)R]]^{1/2}\\-\alpha R[1+(\beta-1)R]\end{array}\right\}}{(1-R)[1+\beta+(\beta-1)R]}.$$
(5.34)

Before proceeding to the various approximations we observe from (5.34) that for all three approximations arising from (5.30) we have, in the limit $t \to t_c$,

$$A_1(t) \sim \left(\frac{2\alpha R}{1+\beta}\right)^{1/2}, \qquad A_2(t) \sim \frac{1}{1+\beta},$$
(5.35)

so that from (5.30) it is apparent that

$$T(r, t_c) = \frac{r}{1+\beta}.$$
(5.36)

Thus the temperature profile at the time of complete freezing is the same for all three approximations and, moreover, the estimate (5.36) satisfies appropriate conditions at both $r = 0$ and $r = 1$. We also note that the following three approximations to the boundary motion arising from (5.30) all coincide with the pseudo-steady-state estimate (1.71) for large α.

5.4.1 Megerlin's method. From the Stefan condition (2.6)$_1$ and (5.30) we have

$$A_1(t) = -\alpha R \frac{dR}{dt},$$
(5.37)

which on using (5.34) becomes

$$2\,dt = -\left\{\{\alpha^2 R^2[1+(\beta-1)R]^2+2\alpha R(1-R)[1+\beta+(\beta-1)R]\}^{1/2} + \alpha R[1+(\beta-1)R]\right\}dR.$$
(5.38)

On integration and using the initial condition $R(0) = 1$ we have

$$t = \frac{\alpha}{12}(1 - R)[(1 + 2\beta)(1 + R) + 2(\beta - 1)R^2]$$

$$+ \frac{1}{2}\int_R^1 \left\{ \alpha^2\xi^2[1 + (\beta - 1)\xi]^2 + 2\alpha\xi(1 - \xi)[1 + \beta + (\beta - 1)\xi] \right\}^{1/2} d\xi,$$
(5.39)

which appears not to be readily expressed in closed form. However, for β zero (5.39) becomes

$$t = \frac{\alpha}{12}(1 - R)^2(1 + 2R) + \frac{1}{2}\int_R^1 (1 - \xi)[\alpha^2\xi^2 + 2\alpha\xi]^{1/2} d\xi,$$
(5.40)

and with the substitution $\eta = 1 + \alpha\xi$ this may be integrated to give

$$t = \frac{\alpha}{12}(1 - R)^2(1 + 2R) + \frac{\alpha + 1}{4\alpha^2}\log\left[\frac{1 + \alpha R + (\alpha^2 R^2 + 2\alpha R)^{1/2}}{1 + \alpha + (\alpha^2 + 2\alpha)^{1/2}}\right]$$

$$+ \frac{1}{12\alpha^2}\left\{ (\alpha^2 R^2 + 2\alpha R)^{1/2}[2\alpha^2 R^2 + (1 - 3\alpha)\alpha R - 3(\alpha + 1)] \right.$$

$$\left. + (\alpha^2 + 2\alpha)^{1/2}(\alpha^2 + 2\alpha + 3) \right\}.$$
(5.41)

Thus we obtain the following estimate for the time t_c to complete freezing:

$$t_c = \frac{\alpha}{12} + \frac{(\alpha^2 + 2\alpha)^{1/2}}{12\alpha^2}(\alpha^2 + 2\alpha + 3) - \frac{\alpha + 1}{4\alpha^2}\log[1 + \alpha + (\alpha^2 + 2\alpha)^{1/2}].$$
(5.42)

5.4.2 Integral formula method.

From (3.23) and (5.30) we have on performing the integration

$$t = \frac{\alpha}{6}(1 - R)[(1 + 2\beta)(1 + R) + 2(\beta - 1)R^2]$$

$$+ \frac{A_1(t)}{6}(1 - R)^2[1 + 2\beta + (\beta - 1)R]$$

$$+ \frac{A_2(t)}{12}(1 - R)^3[1 + 3\beta + (\beta - 1)R],$$
(5.43)

which together with (5.32) and (5.34) gives the only explicit expression for the boundary motion for all α and β. From (5.35) and (5.43) we have the following estimate for the time to complete freezing:

$$t_c = \frac{\alpha}{6}(1 + 2\beta) + \frac{1 + 3\beta}{12(1 + \beta)}. \tag{5.44}$$

This should be contrasted with the estimate obtained by the integral iteration procedure (see Eq. (4.78)).

5.4.3 Heat-balance method for no surface cooling ($\beta = 0$), If we introduce $v(r, t)$ such that

$$T(r, t) = \frac{v(r, t)}{r}, \tag{5.45}$$

then the Stefan problem (2.4)–(2.6) with β zero becomes

$$\frac{\partial v}{\partial t} = \frac{\partial^2 v}{\partial r^2}, \qquad R(t) < r < 1, \tag{5.46}$$

$$v(1, t) = 1, \qquad v(R(t), t) = 0, \tag{5.47}$$

$$\frac{\partial v}{\partial r}(R(t), t) = -\alpha R \frac{dR}{dt}, \qquad R(0) = 1. \tag{5.48}$$

On integrating (5.46) over the interval and using the conditions at the moving boundary we have

$$\frac{d}{dt}\left(\int_R^1 v(\xi, t)\, d\xi - \frac{\alpha R^2}{2}\right) = \frac{\partial v}{\partial r}(1, t), \tag{5.49}$$

which we take to be the heat-balance condition. We observe that a different heat-balance condition might be obtained by integrating (2.4) over the interval (that is, Eq. (2.116)). The two possibilities give rise to distinct boundary motions and the analysis for both of them is reasonably involved. We have taken (5.49) since the calculations are simpler.

From (5.30), (5.45) and (5.49) we have

$$\frac{d}{dt}\left(\frac{A_1(t)}{2}(1 - R)^2 + \frac{A_2(t)}{3}(1 - R)^3 - \frac{\alpha R^2}{2}\right) = [A_1(t) + 2A_2(t)(1 - R)], \tag{5.50}$$

while from (5.34) and (5.32) for β zero we obtain

$$A_1(t) = \frac{(\alpha^2 R^2 + 2\alpha R)^{1/2} - \alpha R}{(1 - R)}, \tag{5.51}$$

$$A_2(t) = \frac{1 + \alpha R - (\alpha^2 R^2 + 2\alpha R)^{1/2}}{(1 - R)^2}. \tag{5.52}$$

From these equations we find that (5.50) becomes

$$\frac{d}{dt}\left(\frac{(1-R)}{6}[2 - \alpha R + (\alpha^2 R^2 + 2\alpha R)^{1/2}] - \frac{\alpha R^2}{2}\right)$$

$$= \frac{[2 + \alpha R - (\alpha^2 R^2 + 2\alpha R)^{1/2}]}{(1 - R)},$$

(5.53)

which with the substitution $y = 1 + \alpha R$ yields

$$\frac{d}{dt}\left(\frac{(\alpha + 1 - y)}{6\alpha}[3 - y + (y^2 - 1)^{1/2}] - \frac{(y - 1)^2}{2\alpha}\right) = \alpha \frac{[1 + y - (y^2 - 1)^{1/2}]}{(\alpha + 1 - y)}.$$

(5.54)

This equation may be rearranged to obtain

$$dt = -\frac{(\alpha + 1 - y)}{12\alpha^2}\left\{\frac{6y^2 + y + \alpha - 3}{(y + 1)} + \frac{6y^2 - 7y + 1 - \alpha}{(y^2 - 1)^{1/2}}\right\} dy,$$

(5.55)

and integrated to eventually give

$$t = \frac{(1 - R)}{24\alpha}\left\{(2\alpha^2 - \alpha - 4) + (2\alpha - 1)\alpha R - 4\alpha^2 R^2\right\} - \frac{(\alpha + 2)^2}{12\alpha^2}$$

$$\times \log\left(\frac{\alpha R + 2}{\alpha + 2}\right) + \frac{1}{24\alpha^2}\left\{[3(2\alpha + 5) - (6\alpha + 5)\alpha R + 4\alpha^2 R^2]\right.$$

$$\left.\times (\alpha^2 R^2 + 2\alpha R)^{1/2} - (15 + \alpha - 2\alpha^2)[\alpha(\alpha + 2)]^{1/2}\right\}$$

$$- \frac{(15 + 6\alpha - 2\alpha^2)}{24\alpha^2}\log\left(\frac{1 + \alpha R + (\alpha^2 R^2 + 2\alpha R)^{1/2}}{1 + \alpha + (\alpha^2 + 2\alpha)^{1/2}}\right).$$

(5.56)

From this equation we obtain the following estimate for the time t_c to complete freezing

$$t_c = \frac{1}{24\alpha^2}\left\{\alpha(2\alpha^2 - \alpha - 4) + (\alpha^2 + 2\alpha)^{1/2}(2\alpha^2 - \alpha - 15) - 2(\alpha + 2)^2\right.$$

$$\left.\times \log\left(\frac{2}{\alpha + 2}\right) - (2\alpha^2 - 6\alpha - 15)\log[1 + \alpha + (\alpha^2 + 2\alpha)^{1/2}]\right\}.$$

(5.57)

Numerical values of the various estimates of t_c are given in Table 5.4 for various values of α. The final column gives the value predicted by a numerical enthalpy scheme (see Voller and Cross 1981) and the asterisk signifies the polynomial approximation which is closest to this value. As far as the three quadratic approximations are concerned numerical results indicate there is no uniformly best estimate. The heat-balance approximation applies for extremely small α, then for $\alpha \geq 0.2$ the integral

Table 5.4 Numerical values of t_c for the sphere for β zero, for various approximations and various values of α

α	Heat balance (quadratic) (5.57)	Integral formula (quadratic) (5.44)	Integral formula (cubic) (5.103)	Megerlin (cubic) (5.101)	Hill and Kucera (1983a) (6.77)	Megerlin (quadratic) (5.42)	Numerical enthalpy scheme (Voller and Cross 1981)
0.1	0.09*	0.10	0.12	0.06	0.10	0.07	0.09
0.2	0.11	0.12*	0.13	0.09	0.13	0.10	0.12
0.5	0.14	0.17	0.18	0.16	0.19	0.18*	0.18
1.0	0.21	0.25	0.27*	0.26	0.29	0.29	0.28
2.0	0.36	0.42	0.43	0.44*	0.47	0.48	0.45
5.0	0.82	0.92	0.93	0.96*	0.99	1.02	0.97
10.0	1.63	1.75	1.77	1.81*	1.83	1.87	1.81
20.0	3.27	3.42	3.43	3.49*	3.50	3.55	3.48

NOTE. The asterisk (*) denotes the polynomial approximation closest to the value predicted by a numerical enthalpy scheme (Voller and Cross 1981)

formula approximation or Megerlin's approximation gives the best estimate. Also given in the table are numerical values arising from both the
cubic profile (5.92) and the boundary-fixing series approximation which is
described in the next chapter. These results are discussed subsequently.

Finally in this section we note the possible alternative quadratic temperature profile

$$T(r, t) = a_1(t)\left(\frac{1}{r} - \frac{1}{R}\right) + a_2(t)\left(\frac{1}{r} - \frac{1}{R}\right)^2, \tag{5.58}$$

and from this equation and (5.31) we may deduce

$$a_2(t) = \frac{a_1(t)^2}{2\alpha}, \tag{5.59}$$

while the surface condition $(2.5)_1$ yields

$$a_1(t)^2(1 - R)[1 + (2\beta - 1)R] - 2\alpha Ra_1(t)[1 + (\beta - 1)R] - 2\alpha R^2 = 0. \tag{5.60}$$

The appropriate solution of (5.60) which ensures $T \geq 0$ is

$$a_1(t) = \frac{\left\{\begin{array}{l} \alpha R[1 + (\beta - 1)R] \\ \quad - \{\alpha^2 R^2[1 + (\beta - 1)R]^2 + 2\alpha R^2(1 - R)[1 + (2\beta - 1)R]\}^{1/2} \end{array}\right\}}{(1 - R)[1 + (2\beta - 1)R]}, \tag{5.61}$$

and thus in the limit as $t \to t_c$ we have

$$a_1(t) \sim [\alpha - (\alpha^2 + 2\alpha)^{1/2}]R, \qquad a_2(t) \sim [1 + \alpha - (\alpha^2 + 2\alpha)^{1/2}]R^2. \tag{5.62}$$

From (5.58) and (5.62) we obtain $T(r, t_c) = 1$, which is physically less
realistic than (5.36) and consequently we do not pursue (5.58) further.

5.5 Freezing a liquid in a long circular cylindrical container

In this section we consider the cylindrical freezing problem (2.7) with
(2.5) and (2.6) with a quadratic temperature profile based on the pseudo-
steady-state solution (1.73), namely

$$T(r, t) = A_1(t)(\log r - \log R) + A_2(t)(\log r - \log R)^2. \tag{5.63}$$

In place of (5.31) we have

$$\left(\frac{\partial T}{\partial r}(R, t)\right)^2 = \alpha\left(\frac{\partial^2 T}{\partial r^2}(R, t) + \frac{1}{R}\frac{\partial T}{\partial r}(R, t)\right), \tag{5.64}$$

from which we have

$$A_2(t) = \frac{A_1(t)^2}{2\alpha}. \tag{5.65}$$

In the usual way we obtain

$$A_1(t) = \frac{[(\alpha^2 + 2\alpha)(\beta - \log R)^2 - 2\alpha\beta^2]^{1/2} - \alpha(\beta - \log R)}{\log R(\log R - 2\beta)}, \tag{5.66}$$

so that in the limit as $t \to t_c$ we have

$$A_1(t) \sim \frac{(\alpha^2 + 2\alpha)^{1/2} - \alpha}{|\log R|}, \qquad A_2(t) \sim \frac{1 + \alpha - (\alpha^2 + 2\alpha)^{1/2}}{|\log R|^2}, \tag{5.67}$$

and therefore the quadratic temperature profile (5.63) is such that $T(r, t_c) = 1$. In the final section of this chapter we consider an alternative approximating temperature profile which is not based on the pseudo-steady-state solution and gives rise to the physically more realistic profile (5.116) at complete freezing.

5.5.1 Megerlin's method.
From the Stefan condition $(2.6)_1$ and (5.63) we obtain (5.37), which using (5.66) can be simplified to give

$$2\,dt = -R\left\{[(\alpha^2 + 2\alpha)(\beta - \log R)^2 - 2\alpha\beta^2]^{1/2} + \alpha(\beta - \log R)\right\}\,dR, \tag{5.68}$$

and on integration we have

$$t = \frac{\alpha}{8}[2R^2\log R + (1 + 2\beta)(1 - R^2)]$$

$$+ \frac{1}{2}\int_R^1 \xi[(\alpha^2 + 2\alpha)(\beta - \log \xi)^2 - 2\alpha\beta^2]^{1/2}\,d\xi. \tag{5.69}$$

For β non-zero the integral appears not to be readily expressed in closed form. However, for β zero (5.69) becomes simply

$$t = \frac{\alpha + (\alpha^2 + 2\alpha)^{1/2}}{8}[2R^2\log R + (1 - R^2)], \tag{5.70}$$

which has the pseudo-steady-state dependence with a different multiplicative constant. For β non-zero we obtain from (5.69) the following estimate for the time t_c to complete solidification:

$$t_c = \frac{\alpha}{8}(1 + 2\beta) + \frac{e^{2\beta}}{2}(\alpha^2 + 2\alpha)^{1/2}\int_\beta^\infty e^{-2\eta}(\eta^2 - \delta^2)^{1/2}\,d\eta, \tag{5.71}$$

where $\delta = \beta(1 + \alpha/2)^{-1/2}$. Again we observe that the integral (5.71) is only readily expressed in closed form for the case when the lower limit of integration is δ.

5.5.2 Integral formula method.

From (3.27) and (5.63) we have on performing the integrations

$$t = \frac{\alpha}{4}[2R^2 \log R + (1+2\beta)(1-R^2)]$$

$$- \frac{A_1(t)}{4}[(1+2\beta+R^2)\log R + (1+\beta)(1-R^2)]$$

$$+ \frac{A_2(t)}{8}[2(1+2\beta)(\log R)^2 + 2(2+2\beta+R^2)\log R + (3+2\beta)(1-R^2)].$$

$$(5.72)$$

In the limit as $t \to t_c$ we obtain from (5.67) and (5.72) simply the upper bound in (3.64), which is to be expected directly from (3.27) since $T(r, t_c) = 1$.

5.5.3 Heat-balance method for no surface cooling $(\beta = 0)$.

On multiplying (2.7) by r, integrating and using (2.6)$_1$ we may show

$$\frac{d}{dt}\left(\int_R^1 \xi T(\xi, t)\, d\xi - \frac{\alpha R^2}{2}\right) = \frac{\partial T}{\partial r}(1, t), \tag{5.73}$$

and using (5.63), (5.65) and (5.66) for β zero we have

$$\int_R^1 \xi T(\xi, t)\, d\xi = \frac{A_2(t) - A_1(t)}{4}[2\log R + (1 - R^2)] + \frac{A_2(t)}{2}(\log R)^2, \tag{5.74}$$

$$\frac{\partial T}{\partial r}(1, t) = A_1(t) - 2A_2(t)\log R, \tag{5.75}$$

$$A_1(t) = \frac{(\alpha^2 + 2\alpha)^{1/2} - \alpha}{|\log R|}, \qquad A_2(t) = \frac{1 + \alpha - (\alpha^2 + 2\alpha)^{1/2}}{|\log R|^2}. \tag{5.76}$$

From these equations we find that (5.73) becomes

$$\frac{d}{dt}\left\{[(\alpha^2 + 2\alpha)^{1/2} - \alpha]\left(2 + \frac{1-R^2}{\log R}\right) + [1 + \alpha - (\alpha^2 + 2\alpha)^{1/2}]\right.$$

$$\left. \times \left(2 + \frac{2}{\log R} + \frac{1-R^2}{(\log R)^2}\right) - 2\alpha R^2\right\} = -4\frac{[2 + \alpha - (\alpha^2 + 2\alpha)^{1/2}]}{\log R},$$

$$(5.77)$$

which on differentiating, integrating and simplifying gives

$$t = \frac{\alpha}{8}\left[1 + \left(\frac{\alpha}{\alpha+2}\right)^{1/2}\right][2R^2\log R + (1 - R^2)] - \frac{1}{2}\left[1 - \left(\frac{\alpha}{\alpha+2}\right)^{1/2}\right]$$

$$+ \frac{1}{4}\left\{\int_R^1 \frac{\xi^2 - 1}{\xi\log\xi}d\xi - \left(\frac{\alpha}{\alpha+2}\right)^{1/2}(1 - R^2) - \left[1 - \left(\frac{\alpha}{\alpha+2}\right)^{1/2}\right]\frac{1 - R^2}{\log R}\right\}.$$

$$(5.78)$$

In the limit as $R \to 0$ the integral in (5.78) is infinite and therefore the heat-balance method with the profile (5.63) predicts an infinite solidification time.

As far as the three quadratic approximations are concerned it is apparent that Megerlin's method predicts the best estimate for t_c, since the integral formula method yields only the known upper bound given in (3.64), while the heat-balance approach predicts an infinite solidification time. However, Megerlin's estimate for t_c although lying between known bounds is not particularly accurate and, clearly, the results of this section are not as useful as those given in the previous section for spherical solidification.

5.6 Cubic approximations for no surface cooling

In this section we consider cubic temperature profiles only for the problems with prescribed surface temperatures (that is, assuming no surface heat loss). Thus, for the problem of planar solidification (2.1)–(2.3) with β zero we seek to approximate the solution by an expression of the form

$$T(x, t) = A_1(t)(x - X) + A_2(t)(x - X)^2 + A_3(t)(x - X)^3. \qquad (5.79)$$

From the surface condition $T(0, t) = 1$ and (5.4) we readily obtain two constraints on the functions $A_1(t)$, $A_2(t)$ and $A_3(t)$:

$$-A_1(t)X + A_2(t)X^2 - A_3(t)X^3 = 1, \qquad A_2(t) = \frac{A_1(t)^2}{2\alpha}. \qquad (5.80)$$

A third equation is obtained by requiring that the partial differential equation (2.1) is satisfied at $x = 0$. On differentiating the surface condition we have

$$\frac{\partial T}{\partial t}(0, t) = 0, \qquad (5.81)$$

and therefore

$$\frac{\partial^2 T}{\partial x^2}(0, t) = 0. \tag{5.82}$$

From (5.79) and (5.82) we readily obtain

$$A_3(t) = \frac{A_2(t)}{3X}. \tag{5.83}$$

Thus, from (5.80) and (5.83) we have

$$A_1(t)^2 X^2 - 3\alpha A_1(t)X - 3\alpha = 0, \tag{5.84}$$

and therefore altogether we find

$$A_1(t) = \frac{3\alpha}{2X}\left[1 - \left(1 + \frac{4}{3\alpha}\right)^{1/2}\right], \quad A_2(t) = \frac{A_1(t)^2}{2\alpha}, \quad A_3(t) = \frac{A_1(t)^2}{6\alpha X}. \tag{5.85}$$

Having specified the temperature profile (5.79) completely in terms of the boundary motion $X(t)$, it remains only to determine this function and two possibilities present themselves, that is, using either the Stefan condition $(2.3)_1$ or the integral formula (3.6).

5.6.1 Megerlin's method. From (5.79) we see that $(2.3)_1$ again gives simply (5.8), which may be readily integrated to yield the usual expression $X(t) = (2\gamma t)^{1/2}$ with the following estimate for γ^{-1}:

$$\gamma^{-1} \simeq \frac{\alpha}{2}\left[1 + \left(1 + \frac{4}{3\alpha}\right)^{1/2}\right]. \tag{5.86}$$

5.6.2 Integral formula method. Alternatively, from (5.79) and the integral formula (3.6) we have

$$t = \frac{X^2}{2}\left\{\alpha - \frac{A_1(t)X}{3} + \frac{A_2(t)X^2}{6} - \frac{A_3(t)X^3}{10}\right\}, \tag{5.87}$$

which on using (5.85) yields the following estimate for γ^{-1}:

$$\gamma^{-1} \simeq \frac{\alpha}{5}\left[4 + \left(1 + \frac{4}{3\alpha}\right)^{1/2}\right] + \frac{1}{5}. \tag{5.88}$$

Numerical values of (5.86) and (5.88) are given in Table 5.2. It is apparent for both quadratic and cubic temperature profiles that use of the integral formula gives the best result.

Although we shall not consider the heat-balance method further, for the sake of completeness we outline Goodman's analysis for planar solidification with β zero and using a cubic temperature profile (see Goodman 1958). For the profile (5.79) Goodman still uses (5.80) but instead of (5.82) employs the heat-balance condition (5.12), which becomes

$$\frac{d}{dt}\left\{\alpha X - \frac{A_1(t)X^2}{2} + \frac{A_2(t)X^3}{3} - \frac{A_3(t)X^4}{4}\right\}$$
$$+ A_1(t) - 2A_2(t)X + 3A_3(t)X^2 = 0. \qquad (5.89)$$

Now, on using the Stefan condition (5.8), Goodman shows that the usual square-root time boundary motion $X(t) = (2\gamma t)^{1/2}$ is obtained. Assuming this to be the case, we may readily deduce from (5.8) and (5.80) the following expressions for $A_1(t)$, $A_2(t)$ and $A_3(t)$:

$$A_1(t) = -\frac{\alpha\gamma}{X}, \qquad A_2(t) = \frac{\alpha\gamma^2}{2X^2}, \qquad A_3(t) = \frac{1}{X^3}\left[\alpha\gamma\left(1 + \frac{\gamma}{2}\right) - 1\right],$$
$$(5.90)$$

and on substituting these expressions into (5.89) and using $X(t) = (2\gamma t)^{1/2}$ we may deduce

$$\frac{1}{\alpha} = \frac{\gamma(12 + \gamma)(6 + \gamma)}{6(12 - \gamma)}, \qquad (5.91)$$

which agrees with Goodman (1958) and provides the equation from which an estimate of γ can be obtained. In fairness to Goodman this solution substantially improves the quadratic heat-balance approximation. However, in view of the complexities involved in the analysis of spherical and cylindrical problems we do not consider the heat-balance approximation further. We close this section with a summary of the main equations for cubic profiles for spherical and cylindrical solidification with β zero.

5.6.3 Spherical freezing.

We approximate the solution of (2.4)–(2.6) with β zero by an expression of the form

$$T(r, t) = \frac{1}{r}\left\{A_1(t)(r - R) + A_2(t)(r - R)^2 + A_3(t)(r - R)^3\right\}. \qquad (5.92)$$

With $v(r, t)$ defined as in (5.45), the three equations for the determination of the three functions $A_1(t)$, $A_2(t)$ and $A_3(t)$ are $v(1, t) = 1$ and

$$\frac{\partial^2 v}{\partial r^2}(1, t) = 0, \qquad \left(\frac{\partial v}{\partial r}(R, t)\right)^2 = \alpha R \frac{\partial^2 v}{\partial r^2}(R, t), \qquad (5.93)$$

from which we may deduce

$$A_1(t) = \frac{3[(\alpha^2 R^2 + \tfrac{4}{3}\alpha R)^{1/2} - \alpha R]}{2(1-R)}, \tag{5.94}$$

$$A_2(t) = \frac{9[(\alpha R + \tfrac{2}{3}) - (\alpha^2 R^2 + \tfrac{4}{3}\alpha R)^{1/2}]}{4(1-R)^2}, \tag{5.95}$$

$$A_3(t) = \frac{3[(\alpha^2 R^2 + \tfrac{4}{3}\alpha R)^{1/2} - (\alpha R + \tfrac{2}{3})]}{4(1-R)^3}. \tag{5.96}$$

Thus, in particular, in the limit as $t \to t_c$ we have

$$A_1(t) \sim (3\alpha R)^{1/2}, \qquad A_2(t) \sim \tfrac{3}{2}, \qquad A_3(t) \sim -\tfrac{1}{2}, \tag{5.97}$$

and therefore from (5.92) we obtain

$$T(r, t_c) = \frac{3r - r^2}{2}, \tag{5.98}$$

which remarkably coincides with the profile at complete solidification predicted by the integral iteration method (see Eq. (4.75) with β zero).

Megerlin's method

The Stefan condition $(2.6)_1$ yields (5.37), which using (5.94) gives

$$2\,dt = -(1-R)[(\alpha^2 R^2 + \tfrac{4}{3}\alpha R)^{1/2} + \alpha R]\,dR \tag{5.99}$$

and on integration we obtain

$$t = \frac{\alpha}{12}(1-R)^2(1+2R) + \frac{3\alpha+2}{27\alpha^2}\log\left[\frac{\alpha R + \tfrac{2}{3} + (\alpha^2 R^2 + \tfrac{4}{3}\alpha R)^{1/2}}{\alpha + \tfrac{2}{3} + (\alpha^2 + \tfrac{4}{3}\alpha)^{1/2}}\right]$$

$$+ \frac{1}{36\alpha^2}\left\{(\alpha^2 R^2 + \tfrac{4}{3}\alpha R)^{1/2}[6\alpha^2 R^2 + (2-9\alpha)\alpha R - 2(2+3\alpha)]\right.$$

$$\left. + (\alpha^2 + \tfrac{4}{3}\alpha)^{1/2}(3\alpha^2 + 4\alpha + 4)\right\}. \tag{5.100}$$

Thus we have the following estimate for the time t_c to complete freezing:

$$t_c = \frac{\alpha}{12} + \frac{(\alpha^2 + \tfrac{4}{3}\alpha)^{1/2}}{36\alpha^2}(3\alpha^2 + 4\alpha + 4)$$

$$- \frac{3\alpha+2}{27\alpha^2}\log\left[\frac{\alpha + \tfrac{2}{3} + (\alpha^2 + \tfrac{4}{3}\alpha)^{1/2}}{\tfrac{2}{3}}\right]. \tag{5.101}$$

Integral formula method

From (3.23) and (5.92) we obtain

$$t = \frac{\alpha}{6}(1-R)^2(1+2R) + \frac{A_1(t)}{6}(1-R)^3 + \frac{A_2(t)}{12}(1-R)^4$$
$$+ \frac{A_3(t)}{20}(1-R)^5, \tag{5.102}$$

which on using (5.94)–(5.96) simplifies to give

$$t = \frac{\alpha}{6}(1-R)^2(1+2R) + \frac{(1-R)^2}{10}[1 - \alpha R + (\alpha^2 R^2 + \tfrac{4}{3}\alpha R)^{1/2}].$$

$$\tag{5.103}$$

We observe that this expression gives $\alpha/6 + 1/10$ for t_c, which is the same as that obtained from the integral iteration method (see Eq. (4.85)). Numerical results given in Table 5.4 indicate that, as far as the two cubic estimates are concerned, Megerlin's method gives slightly superior results to the integral formula method. Moreover, Megerlin's method with a cubic temperature profile emerges as the best estimate out of the various quadratic and cubic approximations.

5.6.4 Cylindrical freezing. We seek to approximate the solution of (2.5), (2.6) and (2.7) with β zero by an expression of this form:

$$T(r, t) = A_1(t)(\log r - \log R) + A_2(t)(\log r - \log R)^2$$
$$+ A_3(t)(\log r - \log R)^3, \tag{5.104}$$

so that from the surface condition $T(1, t) = 1$, (5.64) and

$$\frac{\partial^2 T}{\partial r^2}(1, t) + \frac{\partial T}{\partial r}(1, t) = 0, \tag{5.105}$$

we may readily deduce

$$A_1(t) = \frac{3[(\alpha^2 + \tfrac{4}{3}\alpha)^{1/2} - \alpha]}{2\,|\log R|}, \qquad A_2(t) = \frac{9[\alpha + \tfrac{2}{3} - (\alpha^2 + \tfrac{4}{3}\alpha)^{1/2}]}{4\,|\log R|^2},$$

$$A_3(t) = \frac{3[(\alpha^2 + \tfrac{4}{3}\alpha)^{1/2} - (\alpha + \tfrac{2}{3})]}{4\,|\log R|^3}, \tag{5.106}$$

and again it is apparent that $T(r, t_c) = 1$.

Megerlin's method

From the Stefan condition $(2.6)_1$ and (5.104) we again have (5.37), which yields

$$t = \frac{\alpha + (\alpha^2 + \frac{4}{3}\alpha)^{1/2}}{8}[2R^2 \log R + (1 - R^2)], \tag{5.107}$$

which is again the pseudo-steady-state motion apart from the multiplicative constant.

Integral formula method

From (3.27) with β zero and (5.104) we obtain on performing the integrations

$$t = \frac{\alpha}{4}[2R^2 \log R + (1 - R^2)] - \frac{A_1(t)}{4}[(1 + R^2) \log R + (1 - R^2)]$$

$$+ \frac{A_2(t)}{8}[2(\log R)^2 + 2(2 + R^2) \log R + 3(1 - R^2)]$$

$$- \frac{A_3(t)}{8}[2(\log R)^3 + 6(\log R)^2 + 3(3 + R^2)\log R + 6(1 - R^2)],$$

$$\tag{5.108}$$

which together with (5.106) gives the boundary motion. We may readily confirm that in the limit $t \to t_c$ we again obtain the upper bound expression in (3.64) as the estimate for t_c.

Again we comment that the results arising from the cubic profile (5.104), although lying between known bounds, are less accurate than the corresponding results for spherical solidification arising from (5.92). This is discussed in the final section of the chapter and an alternative approximation to the temperature profile for cylindrical freezing is proposed.

5.7 Alternative temperature profile for cylindrical freezing

Some insight into why the results obtained for the cylindrical problem are less satisfactory than those obtained for either the planar problem or the spherical problem might be gleaned from an examination of the formal series solutions (4.10), (4.18) and (4.29) given in the previous chapter. If we examine the first few terms (carrying out the time differentiations) of either (4.10) or (4.18) then we do indeed obtain polynomial profiles which may be terminated in a natural way to give the approximate temperature profiles employed here. However, if for the cylinder we adopt the same strategy and examine the first few terms of (4.29) then the temperature assumes the form

$$T(r, t) = A_1(t)e_0(z, Z) + A_2(t)e_1(z, Z) + A_3(t)c_1(z, Z) + \ldots,$$

(5.109)

where $z = r^2/4$, $Z = R^2/4$ and Langford's functions $e_0(z, Z)$, $e_1(z, Z)$ and $c_1(z, Z)$ are given explicitly by (2.25) and (2.24), respectively. Although there appears to be no natural way of terminating the series (5.109), it is, however, apparent from these considerations that the temperature profile for the cylindrical problem assumes the general form

$$T(r, t) = (z - Z)F(z, Z) + G(z, Z) \log\left(\frac{z}{Z}\right),$$

(5.110)

where $F(z, Z)$ and $G(z, Z)$ denote arbitrary analytic functions. We observe that (5.110) is linear in $\log(r/R)$. This may explain the shortcomings of (5.63) and (5.104), which evidently contain a quadratic and cubic dependence, respectively, on $\log(r/R)$.

It is clear from (5.110) that an alternative approximating temperature profile for the cylindrical problem might be obtained by replacing $F(z, Z)$ and $G(z, Z)$ in the first instance by functions of Z only, thus

$$T(r, t) = b_1(t)(r^2 - R^2) + b_2(t) \log\left(\frac{r}{R}\right),$$

(5.111)

where $b_1(t)$ and $b_2(t)$ denote functions of time which we determine in the usual way. From the surface condition $(2.5)_1$ and (5.64) we obtain the equations

$$b_1(t)(1 + 2\beta - R^2) + b_2(t)(\beta - \log R) = 1,$$

(5.112)

$$\left[2b_1(t)R + \frac{b_2(t)}{R}\right]^2 = 4\alpha b_1(t),$$

(5.113)

which on solving give

$$b_1(t) = \left\{\frac{[\alpha R^2(\beta - \log R)^2 + 2R^2 \log R + (1 + 2\beta)(1 - R^2)]^{1/2}}{2R^2 \log R + (1 + 2\beta)(1 - R^2)} - \alpha^{1/2}R(\beta - \log R)\right\}^2,$$

(5.114)

and $b_2(t)$ determined from (5.112) and (5.114). In the limit as $t \to t_c$ we have

$$b_1(t) \sim \frac{1}{1 + 2\beta}, \qquad b_2(t) \sim -\frac{R^2}{(1 + 2\beta) \log R},$$

(5.115)

so that the approximating temperature profile at complete freezing is given by

$$T(r, t_c) = \frac{r^2}{1 + 2\beta},$$
(5.116)

which satisfies appropriate conditions at both $r = 0$ and $r = 1$ and as such is physically far more realistic than $T(r, t_c) = 1$.

5.7.1   Megerlin's method.
From the Stefan condition $(2.6)_1$ and the above equations we obtain

$$2\,dt = -\left\{ \alpha R(\beta - \log R) \right.$$
$$\left. + \alpha^{1/2}[\alpha R^2(\beta - \log R)^2 + 2R^2 \log R + (1 + 2\beta)(1 - R^2)]^{1/2} \right\} dR,$$
(5.117)

which on integration gives

$$t = \frac{\alpha}{8}[2R^2 \log R + (1 + 2\beta)(1 - R^2)]$$

$$+ \frac{\alpha^{1/2}}{2} \int_R^1 [\alpha \xi^2 (\beta - \log \xi)^2 + 2\xi^2 \log \xi + (1 + 2\beta)(1 - \xi^2)]^{1/2}\,d\xi.$$
(5.118)

This gives the pseudo-steady-state solution for large α but the integral, even for β zero, appears not to be readily expressed in closed form.

5.7.2   Integral formula method.
From (3.27) and (5.111) we have on performing the integrations

$$t = \frac{\alpha}{4}[2R^2 \log R + (1 + 2\beta)(1 - R^2)]$$

$$+ \frac{b_1(t)}{16}\left\{ [1 + 4\beta - (3 + 4\beta)R^2](1 - R^2) - 4R^4 \log R \right\}$$

$$- \frac{b_2(t)}{4}\left\{ (1 + \beta)(1 - R^2) + (1 + 2\beta + R^2) \log R \right\},$$
(5.119)

which using (5.114) and (5.112) gives an explicit expression for the boundary motion. From the integral formula (3.27) and (5.116) we obtain the following estimate for the time t_c to complete freezing:

$$t_c = \frac{\alpha}{4}(1 + 2\beta) + \frac{1 + 4\beta}{16(1 + 2\beta)}.$$
(5.120)

Although we do not discuss these results in any detail, it is clear that the estimate (5.120) lies between the bounds (3.64).

Finally in this section we note that a possible extension of (5.111) might be

$$T(r, t) = b_1(t)(r^2 - R^2) + [b_2(t) + b_3(t)r^2] \log\left(\frac{r}{R}\right). \tag{5.121}$$

In this case we have

$$\frac{\partial^2 T}{\partial r^2} + \frac{1}{r}\frac{\partial T}{\partial r} = 4\left[b_1(t) + b_3(t) + b_3(t) \log\left(\frac{r}{R}\right)\right], \tag{5.122}$$

and for the case of β zero we may show that the three equations for the determination of $b_1(t)$, $b_2(t)$ and $b_3(t)$ resulting, respectively, from $(2.5)_1$, the time derivative of $(2.5)_1$ and (5.64) are

$$b_1(t)(1 - R^2) - [b_2(t) + b_3(t)] \log R = 1,$$

$$b_1(t) + b_3(t) - b_3(t) \log R = 0, \tag{5.123}$$

$$\left[2b_1(t)R + b_3(t)R + \frac{b_2(t)}{R}\right]^2 = 4\alpha[b_1(t) + b_3(t)].$$

Without solving these equations we may readily confirm that in the limit $t \to t_c$ we have

$$b_1(t) \sim 1 - \log R, \qquad b_2(t) \sim R^2, \qquad b_3(t) \sim -1, \tag{5.124}$$

so that

$$T(r, t_c) = r^2(1 - \log r). \tag{5.125}$$

Again we observe that this profile at complete solidification coincides precisely with that predicted by the integral iteration method (see Eq. (4.92) with β zero) and, moreover, (5.125) together with (3.27) yields $\alpha/4 + 3/32$, which also coincides with (4.95) with β zero.

5.8 Additional symbols used

$a_1(t)$, $a_2(t)$ coefficients in alternative spherical profile, (5.58)

$A_1(t)$, $A_2(t)$, $A_3(t)$ coefficients used for all approximating temperature profiles, (5.2)

b constant $(\alpha^2 + 2\alpha)^{1/2}$, (5.25)

$b_1(t)$, $b_2(t)$, $b_3(t)$ coefficients in alternative cylindrical profiles, (5.111) and (5.121)

$C(t)$, $D(t)$ variables defined by (5.23) and (5.24)

$F(z, Z)$, $G(z, Z)$ arbitrary analytic functions, (5.110)

$v(r, t)$ variable $rT(r, t)$, (5.45)

y	variable $1 + \alpha R$, (5.54)
z, Z	variables $r^2/4$, $R^2/4$, (5.109)

Greek symbols

δ	constant $\beta(1 + \alpha/2)^{-1/2}$, (5.71)
η	integration variable

6

Boundary-fixing transformations and series solutions

6.1 Introduction

In this chapter we utilize boundary-fixing transformations to deduce series solutions of the three basic problems. The boundary-fixing transformations are precisely those employed in Chapter 2 in the derivation of large Stefan number expansions (see Eqs. (2.8)–(2.10)). For convenience we summarize the transformations of basic variables used in Chapter 2 (namely (2.63), (2.79) and (2.94)):

$$\rho = \frac{x}{X(t)}, \qquad \tau = X(t), \qquad T(x, t) = u(\rho, \tau) \quad \text{(plane)}, \tag{6.1}$$

$$\rho = \frac{1-r}{1-R(t)}, \qquad \tau = 1 - R(t), \qquad T(r, t) = \frac{u(\rho, \tau)}{r} \quad \text{(sphere)}, \tag{6.2}$$

$$\rho = \frac{\log r}{\log R(t)}, \qquad \tau = \log R(t), \qquad T(r, t) = u(\rho, \tau) \quad \text{(cylinder)}. \tag{6.3}$$

Here, instead of expanding $u(\rho, \tau)$ in powers of α^{-1}, we look for solutions of the form,

$$u(\rho, \tau) = A_0(\rho) + A_1(\rho)\tau + A_2(\rho)\tau^2 + A_3(\rho)\tau^3 + \dots, \tag{6.4}$$

where $A_n(\rho)$ $(n \geq 0)$ are functions of ρ only. We observe from the above that (6.4) is essentially a small-time solution since, in each case, the variable τ is small initially. For planar solidification the material is of infinite extent, so that we may also consider the large-time solution

$$u(\rho, \tau) = B_0(\rho) + \frac{B_1(\rho)}{(1 + \tau/\beta)} + \frac{B_2(\rho)}{(1 + \tau/\beta)^2} + \frac{B_3(\rho)}{(1 + \tau/\beta)^3} + \dots, \tag{6.5}$$

where again $B_n(\rho)$ $(n \geq 0)$ are functions of ρ only.

In the following section we give solutions of the form (6.4) and (6.5) for the problem of planar solidification with Newton cooling at the surface (that is, β non-zero). For all values of α, the short-time solution (6.4) applies for $X(t) \le \alpha\beta/[(\alpha + 1)(\alpha + 2)]$. However, for large α, (6.4) applies for $X(t)$ well beyond this value. The large-time solution (6.5) gives rise to particularly accurate numerical results for all times apart from the initial period during which (6.4) applies. In Sections 6.3 and 6.4 we consider, respectively, solutions of the form (6.4) for the problems of inward spherical and cylindrical solidification. For these problems the cases β zero and non-zero must be considered separately. For β zero the solution (6.4) gives rise to particularly accurate results for both spherical and cylindrical problems. For purposes of illustration details of the calculations for the determination of $A_1(\rho)$, $A_2(\rho)$ and $A_3'(1)$ are given in Appendix 2 for the problem of spherical solidification with β zero. The solutions for β non-zero, which are less accurate, are included for the sake of completeness and are essentially large β solutions, giving meaningful results provided $\beta \ge 1 + \alpha^{-1}$ for spherical solidification and $\beta \ge 1 + (2\alpha)^{-1}$ for cylindrical solidification. We remark that for β zero the following large α estimates of t_c emerge from the analysis for spherical and cylindrical solidification, respectively,

$$t_c = \frac{\alpha}{6} + \frac{1}{6}, \qquad t_c = \frac{\alpha}{4} + \frac{3}{20}. \tag{6.6}$$

We observe that $(6.6)_1$ is the upper bound in (3.62), while $(6.6)_2$ lies between the bounds (3.64) and is remarkably close to the empirical result

$$t_c = 0.252\alpha + 0.14, \tag{6.7}$$

quoted by Voller and Cross (1981) as being within 1% of their numerical predictions using a finite difference enthalpy method.

We emphasize that the boundary-fixing transformations (6.1)–(6.3) are used, firstly, because these variables are the natural choices associated with the particular geometry and, secondly, because the numerical results are seen to be meaningful and, on occasions, particularly accurate. Although the choice of boundary-fixing transformation for a particular geometry would appear completely arbitrary, it happens that certain transformations are far more effective than others (that is, they give more meaningful numerical results). For example, if Landau's transformation $(6.2)_1$ is utilized for the problem of cylindrical freezing then the ensuing results are completely in error. In order to gain some insight into why the above transformations seem appropriate (or at least the best that have been identified to date), we consider the general problem (1.34) with

(1.35) for $\lambda = 0$ and

$$\rho = \frac{f(r)}{f(R)}, \qquad \tau = f(R), \qquad T(r, t) = u(\rho, \tau), \tag{6.8}$$

where $f(r)$ is as yet arbitrary except that $f(1) = 0$. We find on using the Stefan condition that (1.34) becomes

$$\alpha\left\{\left(\frac{df}{dr}\right)^2 u_{\rho\rho} + \left(\frac{d^2f}{dr^2} + \frac{\lambda}{r}\frac{df}{dr}\right)\tau u_\rho\right\} = \left(\frac{df}{dR}\right)^2 u_\rho(1, \tau)(\rho u_\rho - \tau u_\tau), \tag{6.9}$$

where as usual subscripts denote partial derivatives, and the arguments of u and its partial derivatives are understood to be (ρ, τ) unless otherwise indicated. It is clear that the above boundary-fixing transformations arise as solutions of

$$\frac{d^2f}{dr^2} + \frac{\lambda}{r}\frac{df}{dr} = 0, \qquad f(1) = 0, \tag{6.10}$$

so that

$$f(r) = C\int_r^1 \xi^{-\lambda}\,d\xi, \tag{6.11}$$

where C denotes an arbitrary constant and the integral is $K_\lambda(1, r)$ defined previously by (3.95). Thus, dropping the arbitrary constant we have

$$\rho = \frac{K_\lambda(1, r)}{K_\lambda(1, R)}, \qquad \tau = K_\lambda(1, R), \tag{6.12}$$

and we see that the choice of $f(r)$ appears to produce the greatest simplification in (6.9). We observe that (6.12) certainly summarizes the boundary-fixing transformations (6.1) and (6.3) and strictly speaking (6.2) as well, since the transformations $T = u/r$ reduces the spherical problem to one of planar geometry. However, it is apparent from (6.12) that

$$\rho = \frac{r^{-1} - 1}{R(t)^{-1} - 1}, \qquad \tau = R(t)^{-1} - 1, \qquad T(r, t) = u(\rho, \tau), \tag{6.13}$$

presents a possible alternative for a solution of the spherical problem of the form (6.4). Since τ tends to infinity for times close to complete solidification, (6.4) and (6.13) give a short-time solution, which if we apply an Euler transformation $\tau/(1 + \tau)$ (see, for example, Meksyn 1961) gives $1 - R(t)$ as the appropriate new variable and consequently the transformation (6.2). For the sake of completeness, details for the alternative spherical solution are given in the final section of the chapter for the case of β zero only. We note that although $\log R$ tends to negative infinity as R tends to zero, it is apparent from (6.97) that $A_1(\rho)\tau$ remains finite but it can be shown that $A_2(\rho)\tau^2$ and higher terms do not remain finite. Thus, although a finite number of terms of the temperature profile for the cylindrical problem may become infinite at complete freezing, the solution is justified since it provides an accurate description of the boundary motion which does remain finite. This situation is analogous to the first-order large α expansions given in Chapter 2 which produce singular corrections to the temperature profile but finite contributions to the boundary motion.

Finally in this section we briefly note previous work which relates to the results of this chapter. Firstly, the analysis of Section 6.3 for spherical freezing is essentially equivalent to that noted by Poots (1962b) in section two of his paper, although as pointed out by Bell (1979), Poots's analysis contains a number of typographical errors. Secondly, the series solutions presented by Tao (see Tao (1978) and subsequent work) and discussed in Section 1.5 also relate to the results given here. The essential difference between both Poots (1962b) and Tao (1978) and the present approach is that these authors assume series expansions for both the temperature and unknown boundary in powers of t and $t^{1/2}$, respectively. Here we simply assume the temperature as the series (6.4) in terms of the unknown boundary, which is not determined until the final stages of the analysis. Thus, although the basic solution strategy differs from previous work, the end results when expanded in the same variables must of course coincide. Because of different notations and terminologies the author has not attempted to reconcile these results in detail. However, it is immediately apparent from Poots (1962b) that his results are similar to those given in Section 6.3 and the basic heat functions employed by Tao (1978) and subsequently can be seen also to arise from the present analysis. We have purposely adopted the strategy of not prescribing the form of the moving boundary so that, firstly, the natural form emerges from the analysis rather than an imposed structure and, secondly, we may then, if necessary, investigate the different forms of the boundary motion resulting from either direct use of the Stefan condition or the integral formulae of Chapter 3. The author believes that the determination of exact solutions to these problems (if they exist) hinges on not only leaving the moving boundary arbitrary but also the actual boundary-fixing transformation (that is, to employ a general boundary-fixing transformation such as (6.8)).

6.2 Planar solidification

In this section we consider separately series solutions of the form (6.4) and (6.5) to Eqs. (2.64) and (2.65). These results are due to Hill and Kucera (1985).

6.2.1 Short-time solution.

From (6.4), (2.64) and (2.65) we obtain

$$A_0'' + \gamma \rho A_0' = 0, \tag{6.14}$$

$$A_n'' + \gamma(\rho A_n' - n A_n) = f_n(\rho) \qquad (n \geq 1), \tag{6.15}$$

$$A_0'(0) = 0, \qquad \beta A_1'(0) = A_0(0) - 1, \tag{6.16}$$

$$\beta A_n'(0) = A_{n-1}(0) \quad (n \geq 2), \qquad A_n(1) = 0 \quad (n \geq 0), \tag{6.17}$$

where primes denote differentiation with respect to ρ, γ is given by

$$\gamma = -\frac{A_0'(1)}{\alpha}, \tag{6.18}$$

and the functions $f_n(\rho)$ are defined by

$$f_n(\rho) = \frac{1}{\alpha} \sum_{j=1}^{n} A_j'(1)[\rho A_{n-j}' - (n-j)A_{n-j}] \quad (n \geq 1). \tag{6.19}$$

Equation (6.14) can be readily integrated to yield

$$A_0(\rho) = C_1 + C_2 \int_0^\rho e^{-\gamma \xi^2/2} \, d\xi, \tag{6.20}$$

where C_1 and C_2 denote arbitrary constants. From the n equal to zero boundary conditions it is apparent that these constants are zero and therefore both $A_0(\rho)$ and γ are identically zero in this case. Thus, Eq. (6.15) simplifies considerably and the solutions $A_n(\rho)$ can be obtained simply by integration and using appropriate boundary conditions. The final results for the first six $A_n(\rho)$ are as follows:

$$A_0(\rho) = 0, \qquad A_1(\rho) = \frac{1-\rho}{\beta}, \qquad A_2(\rho) = \frac{\rho-1}{2\alpha\beta^2}\left\{(\rho - 1) + 2(\alpha + 1)\right\},$$

$$A_3(\rho) = \frac{\rho-1}{6\alpha^2\beta^3}\left\{\alpha(\rho - 1)^2 - 6(\alpha + 1)(\rho - 1) - 6(\alpha + 1)(\alpha + 2)\right\},$$

$$A_4(\rho) = \frac{\rho-1}{24\alpha^3\beta^4}\left\{-(3\alpha + 2)(\rho - 1)^3 - 4(\alpha + 1)(3\alpha + 2)(\rho - 1)^2 \right.$$
$$\left. + 12(\alpha + 1)(3\alpha + 5)(\rho - 1) + 8(\alpha + 1)(3\alpha^2 + 16\alpha + 17)\right\},$$

$$A_5(\rho) = \frac{\rho-1}{120\alpha^4\beta^5}\left\{-\alpha(3\alpha + 2)(\rho - 1)^4 + 10(\alpha + 1)(6\alpha + 7)(\rho - 1)^3 \right.$$
$$+ 20(\alpha + 1)(\alpha + 2)(6\alpha + 7)(\rho - 1)^2$$
$$- 40(\alpha + 1)(6\alpha^2 + 25\alpha + 23)(\rho - 1)$$
$$\left. - 40(\alpha + 1)(3\alpha^3 + 31\alpha^2 + 81\alpha + 60)\right\}. \tag{6.21}$$

We observe that the pseudo-steady-state estimate (1.66) emerges from (6.4) and these expressions, simply by retaining the order one contribution and neglecting terms of order α^{-1} and higher powers, thus

$$u(\rho, \tau) = (1 - \rho)\frac{X}{\beta}\left\{1 - \left(\frac{X}{\beta}\right) + \left(\frac{X}{\beta}\right)^2 - \left(\frac{X}{\beta}\right)^3 + \left(\frac{X}{\beta}\right)^4 + \ldots\right\}$$
$$+ O\left(\frac{1}{\alpha}\right), \tag{6.22}$$

which, on assuming the series in (6.22) is the geometric series, gives precisely (1.66).

The motion of the boundary can now be determined either from the Stefan condition (2.66) or from the integral formula (3.6). For the first method we have

$$dt = -\alpha\{A_1'(1) + A_2'(1)\tau + A_3'(1)\tau^2 + \ldots\}^{-1}\,d\tau, \tag{6.23}$$

which we may expand correct to order τ^4 so that after a long but straightforward calculation we obtain

$$t = \alpha\beta X$$
$$+ (\alpha + 1)X^2\left\{\frac{1}{2} - \frac{X}{3\alpha\beta} + \frac{(\alpha + 2)X^2}{3\alpha^2\beta^2} - \frac{(5\alpha^2 + 27\alpha + 29)X^3}{15\alpha^3\beta^3} + \ldots\right\}, \tag{6.24}$$

which is correct up to order X^5. Using the above equations with the integral formula (3.6) confirms (6.24) and, moreover, gives the motion correct up to order X^6. On retaining terms of order α and order one in (6.24), that is

$$t = \frac{\alpha X}{2}(X + 2\beta) + \frac{X^2}{2} - \frac{X^3}{3\beta}\left\{1 - \left(\frac{X}{\beta}\right) + \left(\frac{X}{\beta}\right)^2 + \ldots\right\} + O\left(\frac{1}{\alpha}\right), \tag{6.25}$$

we see that the order one corrected motion given by (2.41) emerges, assuming the series in (6.25) is geometric with sum $(1 + X/\beta)^{-1}$.

Numerical values of (6.24) indicate that the motion is meaningful only for sufficiently small values of X. For example, for all values of α the condition

$$X(t) \le \frac{\alpha\beta}{(\alpha + 1)(\alpha + 2)}, \tag{6.26}$$

is sufficient to ensure that (6.24) lies between the bounds (3.59). For the lower bound we require to show

$$\frac{(\alpha + 1)X}{3\alpha\beta}\left\{1 + \frac{(5\alpha^2 + 27\alpha + 29)X^2}{5\alpha^2\beta^2}\right\} \le \frac{1}{2} + \frac{(\alpha + 1)(\alpha + 2)X^2}{3\alpha^2\beta^2}, \tag{6.27}$$

if (6.26) holds. Consider the left-hand side

$$\frac{(\alpha + 1)X}{3\alpha\beta}\left\{1 + \frac{(5\alpha^2 + 27\alpha + 29)X^2}{5\alpha^2\beta^2}\right\}$$

$$\leq \frac{1}{3(\alpha + 2)}\left\{1 + \frac{5\alpha^2 + 27\alpha + 29}{5(\alpha + 1)^2(\alpha + 2)^2}\right\}$$

$$= \frac{1}{3(\alpha + 2)}\left\{1 + \frac{1}{(\alpha + 1)(\alpha + 2)} + \frac{12\alpha + 19}{5(\alpha + 1)^2(\alpha + 2)^2}\right\}$$

$$\leq \frac{1}{3(\alpha + 2)}\left\{1 + \frac{1}{(\alpha + 1)(\alpha + 2)} + \frac{12}{5(\alpha + 1)^2(\alpha + 2)}\right\}$$

$$\leq \frac{1}{6}\left(1 + \frac{1}{2} + \frac{6}{5}\right) = \frac{9}{20}$$

$$\leq \frac{1}{2} + \frac{(\alpha + 1)(\alpha + 2)X^2}{3\alpha^2\beta^2},$$

so that the pseudo-steady-state motion is certainly a lower bound. For the upper bound we require to show

$$1 \leq \frac{(X + \beta)(\alpha + 1)}{\alpha\beta}\left\{1 - \frac{(\alpha + 2)X}{\alpha\beta} + \frac{(5\alpha^2 + 27\alpha + 29)X^2}{5\alpha^2\beta^2}\right\}, \qquad (6.28)$$

if (6.26) holds. Consider the right-hand side

$$\frac{(X + \beta)(\alpha + 1)}{\alpha\beta}\left\{1 - \frac{(\alpha + 2)X}{\alpha\beta} + \frac{(5\alpha^2 + 27\alpha + 29)X^2}{5\alpha^2\beta^2}\right\}$$

$$\geq \frac{(X + \beta)(\alpha + 1)}{\alpha\beta}\left\{1 - \frac{(\alpha + 2)X}{\alpha\beta} + \frac{(\alpha + 2)^2X^2}{\alpha^2\beta^2}\right\}$$

$$= \frac{\alpha + 1}{\alpha}\left\{1 - \frac{2X}{\alpha\beta} + \frac{2(\alpha + 2)X^2}{\alpha^2\beta^2} + \frac{(\alpha + 2)^2X^3}{\alpha^2\beta^3}\right\}$$

$$\geq \frac{\alpha + 1}{\alpha}\left\{1 - \frac{2X}{\alpha\beta}\right\} \geq \frac{(\alpha + 1)}{\alpha}\left\{1 - \frac{2}{(\alpha + 1)(\alpha + 2)}\right\}$$

$$= \left(\frac{\alpha + 3}{\alpha + 2}\right) \geq 1,$$

so that the order one corrected motion is certainly an upper bound. We note that for large α, (6.24) is valid beyond the range (6.26). Since the above analysis provides only a strict short-time solution we do not pursue these results further. The following large-time solution provides a more useful estimate.

6.2.2 Large-time solution. With variables defined by

$$\rho = \frac{x}{X(t)}, \qquad \sigma = \frac{1}{1 + X(t)/\beta}, \qquad T(x, t) = v(\rho, \sigma), \tag{6.29}$$

the problem (2.1)–(2.3) becomes

$$\alpha v_{\rho\rho} = v_\rho(1, \sigma)[\rho v_\rho + \sigma(1 - \sigma)v_\sigma], \tag{6.30}$$

$$\sigma v_\rho(0, \sigma) = (1 - \sigma)[v(0, \sigma) - 1], \qquad v(1, \sigma) = 0, \tag{6.31}$$

$$v_\rho(1, \sigma) = \alpha\beta^2 \frac{(1 - \sigma)}{\sigma^3} \frac{d\sigma}{dt}, \qquad \sigma(0) = 1, \tag{6.32}$$

and the large-time solution (6.5) becomes

$$v(\rho, \sigma) = B_0(\rho) + B_1(\rho)\sigma + B_2(\rho)\sigma^2 + B_3(\rho)\sigma^3 + \dots . \tag{6.33}$$

On substituting (6.33) into (6.30) we obtain

$$B_0'' + \gamma\rho B_0' = 0, \tag{6.34}$$

$$B_n'' + \gamma(\rho B_n' + nB_n) = g_n(\rho) \quad (n \geq 1), \tag{6.35}$$

where again $\gamma = -B_0'(1)/\alpha$ and the functions $g_n(\rho)$ are defined by

$$g_n(\rho) = (n - 1)\gamma B_{n-1}$$
$$+ \frac{1}{\alpha}\sum_{j=1}^{n} B_j'(1)[\rho B_{n-j}' + (n - j)B_{n-j} - (n - j - 1)B_{n-j-1}] \tag{6.36}$$

with the convention that B_{-1} is taken to be zero. From (6.31) and (6.33) we have

$$B_0(0) = 1, \qquad B_1(0) = B_0'(0), \tag{6.37}$$

$$B_n(0) = B_{n-1}'(0) + B_{n-1}(0) \quad (n \geq 2), \qquad B_n(1) = 0 \quad (n \geq 0). \tag{6.38}$$

For n zero we readily obtain from the above equations

$$B_0(\rho) = \alpha\gamma \int_\rho^1 e^{\gamma(1 - \xi^2)/2} \, d\xi, \tag{6.39}$$

where γ is determined as a positive root of the transcendental equation

$$\alpha\gamma \int_0^1 e^{\gamma(1 - \xi^2)/2} \, d\xi = 1, \tag{6.40}$$

and these evidently coincide with (1.49) and (1.43), respectively. Thus the leading term of (6.33) is essentially Neumann's exact solution, which is consequently recovered from the present analysis in the limit β tending to zero. For $n \geq 1$ we may deduce the following expression for $B_n(\rho)$:

$$B_n(\rho) = \int_\rho^1 g_n(\xi)[B_{n1}(\rho)B_{n2}(\xi) - B_{n1}(\xi)B_{n2}(\rho)]\, e^{\gamma\xi^2/2}\, d\xi$$
$$+ B_n'(1)B_{n1}(1)\, e^{\gamma/2}B_{n2}(\rho), \tag{6.41}$$

where $B_{n1}(\rho)$ and $B_{n2}(\rho)$ are linearly independent solutions of the homogeneous equation in (6.35) such that

$$B_{n1}(\rho) = \frac{d^n}{d\rho^n}\left(\int_\rho^1 e^{-\gamma\xi^2/2}\, d\xi\right),$$
$$B_{n2}(\rho) = -B_{n1}(\rho)\int_\rho^1 \frac{e^{-\gamma\xi^2/2}}{B_{n1}(\xi)^2}\, d\xi, \tag{6.42}$$

that is their Wronskian $\omega(\rho)$ is given by

$$\omega(\rho) = B_{n2}B_{n1}' - B_{n1}B_{n2}' = -e^{-\gamma\rho^2/2}. \tag{6.43}$$

For example, the first three functions $B_{n1}(\rho)$ are given by

$$B_{11}(\rho) = -e^{-\gamma\rho^2/2}, \qquad B_{21}(\rho) = \gamma\rho\, e^{-\gamma\rho^2/2},$$
$$B_{31}(\rho) = \gamma(1 - \gamma\rho^2)\, e^{-\gamma\rho^2/2}. \tag{6.44}$$

With $B_n(\rho)$ given by (6.41) the determining equation for the unknown $B_n'(1)$ results from (6.41) and the boundary conditions $(6.37)_2$ and $(6.38)_1$ (details of a similar calculation are given in Appendix 2 for the problem of spherical solidification).

After a long but straightforward calculation the final expressions for $n = 1$, 2 and 3 are

$$B_1(\rho) = -\alpha\gamma(1 - \rho)\, e^{\gamma(1-\rho^2)/2},$$

$$B_2(\rho) = \frac{\alpha\gamma^2}{2}\rho(1 - \rho)^2\, e^{\gamma(1-\rho^2)/2},$$

$$B_3(\rho) = \frac{\alpha\gamma^2}{6}\, e^{\gamma(1-\rho^2)/2}\left\{(1 - \gamma\rho^2)(1 - \rho)^3\right.$$
$$\left. + 2\left[\rho(e^{\gamma/2} - e^{\gamma\rho^2/2}) + (1 - \gamma\rho^2)\int_\rho^1 e^{\gamma\xi^2/2}\, d\xi\right]\left(\int_0^1 e^{\gamma\xi^2/2}\, d\xi\right)^{-1}\right\}. \tag{6.45}$$

We make the following two observations concerning the above derivations. Firstly, for $B_2(\rho)$ we see from (6.36), $(6.42)_2$ and $(6.44)_2$ that the term involving $B_2'(1)$ in (6.41) has a logarithmic singularity unless $B_2'(1)$ is zero. Remarkably, the remaining terms in $B_2(\rho)$ accommodates this requirement. Secondly, the identity

$$g_3(\rho) = -\frac{\alpha\gamma^2}{2(1-\gamma\rho^2)}\frac{d}{d\rho}\left\{(1-\rho)^2(1-\gamma\rho^2)^2\,e^{\gamma(1-\rho^2)/2}\right\}$$
$$- \gamma B_3'(1)\rho\,e^{\gamma(1-\rho^2)/2}, \tag{6.46}$$

is an important equation for simplifying (6.41) for $n = 3$.

From (6.39) and (6.45) we have the following expressions for $B_n'(1)$:

$$B_0'(1) = -\alpha\gamma, \qquad B_1'(1) = \alpha\gamma, \qquad B_2'(1) = 0,$$

$$B_3'(1) = -\alpha\gamma^2\,e^{\gamma/2}\left(3\int_0^1 e^{\gamma\xi^2/2}\,d\xi\right)^{-1}, \tag{6.47}$$

and from the Stefan condition (6.32) we find on integration that the motion of the boundary becomes

$$t = \frac{X}{2\gamma}(X+2\beta) - \frac{\beta^2 X\,e^{\gamma/2}}{3(X+\beta)}\left(\int_0^1 e^{\gamma\xi^2/2}\,d\xi\right)^{-1} + \dots. \tag{6.48}$$

We note that the significance of $B_2'(1)$ zero is that there results no $\log\sigma$ contribution on integrating (6.32). On expanding (6.48) for large α and using (1.58) we may show that precisely the order one corrected motion (2.41) emerges. Numerical results indicate that the large-time motion (6.48) lies between the improved upper and lower bounds (3.66) and (3.76), respectively. Table 6.1 gives numerical values of $t/\alpha\beta^2$ from

Table 6.1 Numerical values of $t/\alpha\beta^2$ as given by (6.48) and (2.78) for twelve values of X/β used by Pedroso and Domoto (1973a)

X/β	$t/\alpha\beta^2\ (\alpha=1)$		$t/\alpha\beta^2\ (\alpha=2)$		$t/\alpha\beta^2\ (\alpha=3)$	
	(6.48)	(2.78)	(6.48)	(2.78)	(6.48)	(2.78)
0.2	0.22	0.24	0.22	0.23	0.22	0.23
0.4	0.50	0.54	0.50	0.51	0.50	0.50
0.6	0.85	0.90	0.83	0.84	0.82	0.82
0.8	1.27	1.32	1.21	1.23	1.19	1.19
1.0	1.74	1.80	1.64	1.66	1.60	1.61
1.4	2.85	2.91	2.64	2.66	2.56	2.57
1.8	4.17	4.24	3.83	3.85	3.71	3.72
2.2	5.71	5.78	5.22	5.23	5.03	5.04
2.6	7.47	7.54	6.78	6.80	6.53	6.54
3.0	9.43	9.50	8.54	8.56	8.21	8.22
4.0	15.26	15.33	13.73	13.75	13.18	13.19
5.0	22.40	22.46	20.09	20.11	19.27	19.27

Table 6.2 Numerical values of temperature from (6.5) taking two, three and four terms and compared with an implicit enthalpy scheme for $\alpha = 1$ and four boundary positions

$X(t)/\beta$	$x/X(t)$	Two terms of (6.5)	Three terms of (6.5)	Four terms of (6.5)	Numerical enthalpy scheme
1.0	0.2	0.3304	0.3441	0.3842	0.3652
	0.4	0.2387	0.2534	0.2806	0.2656
	0.6	0.1556	0.1647	0.1810	0.1713
	0.8	0.0771	0.0798	0.0870	0.0827
2.0	0.2	0.4787	0.4847	0.4966	0.4928
	0.4	0.3449	0.3514	0.3595	0.3564
	0.6	0.2212	0.2252	0.2301	0.2280
	0.8	0.1065	0.1077	0.1099	0.1089
3.0	0.2	0.5528	0.5562	0.5613	0.5600
	0.4	0.3980	0.4017	0.4051	0.4041
	0.6	0.2540	0.2562	0.2583	0.2576
	0.8	0.1212	0.1219	0.1228	0.1225
4.0	0.2	0.5973	0.5995	0.6021	0.6015
	0.4	0.4299	0.4322	0.4340	0.4335
	0.6	0.2737	0.2751	0.2761	0.2758
	0.8	0.1300	0.1305	0.1309	0.1308

(6.48) and (2.78) for the twelve values of X/β used by Pedroso and Domoto (1973a) and for three values of α. It is apparent from the table that (6.48) gives an accurate estimate of the motion of the boundary. Table 6.2 gives numerical values of the temperature as obtained from the series (6.5) taking two, three and four terms, respectively, for $\alpha = 1$ and four boundary positions. The final column gives the temperature as obtained from an implicit enthalpy scheme using the Voller and Cross (1981) method to track the boundary. It is apparent from this table that the convergence of (6.5) appears satisfactory and, moreover, the temperature predicted compares favourably with the purely numerical result. We comment that for both increasing α and boundary position the convergence of (6.5) and agreement with the numerical prediction is even closer, as would be expected.

We note that using only the first two terms of (6.33) together with the integral formula (3.6) gives an expression of the same form as (6.48) but with a different multiplicative constant in the second term, thus

$$t = \frac{X}{2\gamma}(X + 2\beta) - \frac{\beta^2 X(1 + \alpha + \gamma^{-1} - 2\alpha\, e^{\gamma/2})}{(X + \beta)} + \ldots, \qquad (6.49)$$

and for large α this expression also gives the order one corrected motion. Finally in this section we remark that by expanding $(1 + \tau/\beta)^{-n}$ in (6.5) in powers of τ, using the elementary identity

$$\binom{-n}{k} = (-1)^k \binom{n+k-1}{n-1}, \tag{6.50}$$

and equating to (6.4), we may establish the identity

$$A_n(\rho) = \frac{(-1)^n}{\beta^n} \sum_{j=0}^{\infty} \binom{n+j-1}{j-1} B_j(\rho), \tag{6.51}$$

and we comment that the inverse relations of (6.51) are not immediately apparent.

6.3 Freezing a liquid in a spherical container

In this section we consider solutions of the form (6.4) to the problem of spherical solidification. As noted previously the cases β zero and non-zero must be treated separately. The analysis for β zero is based on Davis and Hill (1982) and Hill and Kucera (1983a), while that for β non-zero is due to Hill and Kucera (1983b). In terms of the variables ρ, τ and $u(\rho, \tau)$ defined by (6.2), the problem of spherical solidification reduces to (2.80) and (2.81) together with (2.82). On substituting (6.4) into (2.80) we again obtain a system of equations of the form (6.14) and (6.15) with γ given by (6.18) and the functions $f_n(\rho)$ defined by

$$f_n(\rho) = A''_{n-1} + \frac{1}{\alpha} \sum_{j=1}^{n} A'_j(1)[\rho A'_{n-j} - (n-j)A_{n-j}] \quad (n \geq 1). \tag{6.52}$$

From (2.81) and (6.4) we find that the boundary conditions for $A_n(\rho)$ for β zero become

$$A_0(0) = 1, \qquad A_0(1) = 0, \tag{6.53}$$

$$A_n(0) = 0, \qquad A_n(1) = 0 \quad (n \geq 1), \tag{6.54}$$

while if β is non-zero we have

$$A'_0(0) = 0, \qquad A_0(1) = 0, \tag{6.55}$$

$$(1 - \beta)A_0(0) - \beta A'_1(0) = 1, \qquad A_1(1) = 0, \tag{6.56}$$

$$\beta A'_n(0) = (1 - \beta)A_{n-1}(0), \qquad A_n(1) = 0 \quad (n \geq 2). \tag{6.57}$$

Accordingly we consider the two cases separately.

6.3.1 Solution for prescribed surface temperature ($\beta = 0$). From

(6.20) and the boundary conditions (6.53) we may readily show that $A_0(\rho)$ is also given by (6.39) with γ again determined as a positive root of (6.40). For $n \geq 1$ the solutions of the homogeneous differential equation in (6.15) are obtained as follows. We let $A_n(\rho) = A(y)$, where $y = -\gamma\rho^2/2$, so that we have

$$y\frac{d^2A}{dy^2} + (\tfrac{1}{2} - y)\frac{dA}{dy} + \frac{n}{2}A = 0, \tag{6.58}$$

which is the confluent hypergeometric equation with linearly independent solutions

$$A_{n1}(\rho) = {}_1F_1\left(\frac{-n}{2}, \frac{1}{2}; \frac{-\gamma\rho^2}{2}\right),$$

$$A_{n2}(\rho) = \rho {}_1F_1\left(\frac{1}{2} - \frac{n}{2}, \frac{3}{2}; \frac{-\gamma\rho^2}{2}\right), \tag{6.59}$$

where ${}_1F_1\,(a, c; y)$ is defined by

$${}_1F_1\,(a, c; y) = \sum_{k=0}^{\infty} \frac{(a)_k\, y^k}{(c)_k\, k!}, \tag{6.60}$$

where the symbol $(a)_k$ is given by

$$(a)_0 = 1, \qquad (a)_k = a(a+1)(a+2)\ldots(a+k-1) \quad (k \geq 1). \tag{6.61}$$

From the differential equation (6.58) and the first terms of these series, we can show that the linearly independent solutions (6.59) have Wronskian $\omega(\rho)$ given by (6.43). We remark that the above solutions are essentially the standard iterated integrals of the complementary error function and are related to the heat functions utilized by Tao (1978) and subsequently. Since this correspondence is not relevant here we do not pursue the matter. For our purposes it is convenient merely to identify appropriate linearly independent solutions of (6.58), so that by the method of variation of parameters we obtain the formal solution of (6.15) as

$$A_n(\rho) = \int_0^\rho \frac{f_n(\xi)}{\omega(\xi)}\left\{A_{n1}(\rho)A_{n2}(\xi) - A_{n1}(\xi)A_{n2}(\rho)\right\}d\xi$$

$$+ C_{n1}A_{n1}(\rho) + C_{n2}A_{n2}(\rho), \tag{6.62}$$

where C_{n1} and C_{n2} denote arbitrary constants. From the boundary conditions (6.54) and the solutions (6.59) we find that C_{n1} is zero while C_{n2} is given by

$$C_{n2} = \int_0^1 \frac{f_n(\xi)}{\omega(\xi)}\left\{A_{n1}(\xi) - \frac{A_{n1}(1)}{A_{n2}(1)}A_{n2}(\xi)\right\}d\xi. \tag{6.63}$$

On differentiating (6.62), setting $\rho = 1$ and using (6.63) we obtain the following expression for $A_n'(1)$:

$$A_n'(1) = \int_0^1 f_n(\xi) \frac{\omega(1)}{\omega(\xi)} \frac{A_{n2}(\xi)}{A_{n2}(1)} \, d\xi. \tag{6.64}$$

Since $A_n'(1)$ appears on both sides of (6.64), it is the basic equation for the determination of $A_n'(1)$.

The first three linearly independent solutions from (6.59) are

$$A_{12}(\rho) = \rho, \qquad A_{11}(\rho) = e^{-\gamma\rho^2/2} + \gamma\rho \int_0^\rho e^{-\gamma\xi^2/2} \, d\xi,$$

$$A_{21}(\rho) = 1 + \gamma\rho^2,$$

$$A_{22}(\rho) = \frac{1}{2}\left\{ \rho \, e^{-\gamma\rho^2/2} + (1 + \gamma\rho^2) \int_0^\rho e^{-\gamma\xi^2/2} \, d\xi \right\}, \tag{6.65}$$

$$A_{32}(\rho) = \rho\left(1 + \frac{\gamma\rho^2}{3}\right),$$

$$A_{31}(\rho) = \frac{1}{2}\left\{ (2 + \gamma\rho^2) \, e^{-\gamma\rho^2/2} + \gamma\rho(3 + \gamma\rho^2) \int_0^\rho e^{-\gamma\xi^2/2} \, d\xi \right\}.$$

From the above equations for $n = 1$ and $n = 2$ we may deduce, after a long but straightforward calculation (the details of which are given in Appendix 2), the following expressions for $A_1(\rho)$ and $A_2(\rho)$:

$$A_1(\rho) = \frac{\alpha\gamma\rho}{(\gamma + 3)}\left\{ 1 - e^{\gamma(1-\rho^2)/2} \right\},$$

$$A_2(\rho) = \frac{\alpha\gamma}{(\gamma + 3)^2}\left\{ \alpha\gamma\mu(1 + \gamma\rho^2) \int_0^\rho e^{\gamma(1-\xi^2)/2} \, d\xi \right. \tag{6.66}$$

$$\left. - \left[\mu(\gamma + 1) + \frac{\gamma}{2}(1 - \rho^2) \right] \rho \, e^{\gamma(1-\rho^2)/2} \right\},$$

where the constant μ is given by

$$\mu = 6[(\gamma^2 + 6\gamma + 3) + \alpha\gamma(\gamma + 5)]^{-1}. \tag{6.67}$$

Although an expression for $A_3(\rho)$ has not been found, details for the determination of $A_3'(1)$, due to Hill and Kucera (1983a), can be found in Appendix 2. Altogether from the above equations and Appendix 2 we have

$$A_0'(1) = -\alpha\gamma, \qquad A_1'(1) = \frac{\alpha\gamma^2}{(\gamma + 3)}, \qquad A_2'(1) = \frac{\alpha\gamma\mu_1}{(\gamma + 3)^2},$$

$$A_3'(1) = \frac{\alpha\gamma\mu_2}{(\gamma + 3)^3}, \tag{6.68}$$

where the constants μ_1 and μ_2 are defined by

$$\mu_1 = \left\{ \frac{(\gamma^3 + 12\gamma^2 + 15\gamma - 6) + \alpha\gamma(\gamma^2 + 11\gamma + 6)}{(\gamma^2 + 6\gamma + 3) + \alpha\gamma(\gamma + 5)} \right\},$$

$$\mu_2 = \frac{\mu_1}{8} \left\{ \frac{(2\gamma^4 - 14\gamma^3 - 121\gamma^2 - 170\gamma + 135) + \alpha\gamma(2\gamma^3 - 16\gamma^2 - 105\gamma - 135)}{(\gamma^2 + 10\gamma + 15)(\gamma + 1 + \alpha\gamma)} \right\}$$

$$+ \frac{1}{24} \left\{ \frac{\begin{array}{c}(2\gamma^5 + 182\gamma^4 + 1575\gamma^3 + 4590\gamma^2 + 3915\gamma - 2430) \\ + \alpha\gamma(2\gamma^4 + 180\gamma^3 + 1395\gamma^2 + 3375\gamma + 2430)\end{array}}{(\gamma^2 + 10\gamma + 15)(\gamma + 1 + \alpha\gamma)} \right\}.$$

$$(6.69)$$

In order to determine the motion of the boundary we have from (2.82) and (6.4)

$$dt = -\alpha\tau(1 - \tau)\{A_0'(1) + A_1'(1)\tau + A_2'(1)\tau^2 + A_3'(1)\tau^3 + \ldots\}^{-1} d\tau,$$

$$(6.70)$$

which, on expanding in powers of τ and integrating, gives on simplification

$$t = \frac{1}{\gamma} \left\{ \frac{(1 - R)^2}{2} - \frac{(1 - R)^3}{(\gamma + 3)} + \frac{\lambda_1(1 - R)^4}{4(\gamma + 3)^2} + \frac{\lambda_2(1 - R)^5}{5(\gamma + 3)^3} + \ldots \right\}, \quad (6.71)$$

where the constants λ_1 and λ_2 are given by

$$\lambda_1 = \mu_1 - 3\gamma, \qquad \lambda_2 = \mu_2 + (\gamma - 3)\mu_1 - 3\gamma^2. \quad (6.72)$$

We remark that some of the above formulae differ slightly from those given in Davis and Hill (1982) and Hill and Kucera (1983a) since the basic expansion variable used there is $R - 1$ in contrast to $1 - R$ employed here. A remarkable characteristic of this solution is that the above determination of the motion is precisely equivalent to the determination via the integral formula (3.23) with β zero. In this case from

$$T(r, t) = \frac{1}{r} \sum_{n=0}^{\infty} A_n(\rho)(1 - R)^n, \quad (6.73)$$

and (3.23) we have

$$t = t_{\text{pss}}(R) + \sum_{n=0}^{\infty} I_n(1 - R)^{n+2}, \quad (6.74)$$

where $t_{\text{pss}}(R)$ is given by (1.71) with β zero and the constants I_n are defined by

$$I_n = \int_0^1 \rho A_n(\rho) \, d\rho. \tag{6.75}$$

For example, with just two terms of the series we have

$$I_0 = \frac{1}{2\gamma} - \frac{\alpha}{2}, \qquad I_1 = \frac{\alpha}{3} - \frac{1}{\gamma(\gamma + 3)}, \tag{6.76}$$

and (6.74) gives precisely the first two terms of (6.71). Since the motion of the boundary from the Stefan condition coincides with that from the integral formula, this means that the assumed form of the temperature profile (6.4) or (6.73) is exactly appropriate in this case.

From (6.71) we obtain the following estimate of the time to complete solidification

$$t_c = \frac{1}{\gamma} \left\{ \frac{1}{2} - \frac{1}{(\gamma + 3)} + \frac{\lambda_1}{4(\gamma + 3)^2} + \frac{\lambda_2}{5(\gamma + 3)^3} + \cdots \right\}. \tag{6.77}$$

Now for large α (small γ) we have from (6.69) using the first two terms of (1.57)

$$\mu_1 = 3\gamma + O(\gamma^2), \qquad \mu_2 = 9\gamma + O(\gamma^2), \tag{6.78}$$

so that λ_1 and λ_2 are both of order γ^2 and Eq. (6.77) yields

$$t_c = \frac{1}{\gamma} \left(\frac{1}{6} + \frac{\gamma}{9} + O(\gamma^2) \right) = \frac{\alpha}{6} + \frac{1}{6} + O\left(\frac{1}{\alpha}\right), \tag{6.79}$$

which is the correct large α estimate. For small α (large γ) we have from (1.53)

$$\gamma = 2 \log(\alpha^{-1}) + O[\log \log(\alpha^{-1})], \tag{6.80}$$

while from (6.77) we obtain

$$t_c = \frac{1}{2\gamma} + O\left(\frac{1}{\gamma^2}\right), \tag{6.81}$$

and Eqs. (6.80) and (6.81) yield the asymptotically correct small α estimate of t_c.

Table 5.4 gives numerical values of t_c as obtained from (6.77). The results are seen to be uniformly close to the values predicted by the numerical enthalpy scheme of Voller and Cross (1981) which are given in the final column of the table. In fact, since (6.4) is a small-time solution, the agreement of (6.71) with results from the numerical enthalpy scheme is considerably better for intermediate boundary positions. In order to indicate the convergence of (6.71), numerical values of the successive

Table 6.3 Numerical values of successive estimates $t_1(R)$, $t_2(R)$ and $t_3(R)$ to spherical boundary motion for three values of α and five equally spaced boundary positions and compared with results from the numerical enthalpy scheme of Voller and Cross (1981)

Value of α	Boundary position $R(t)$	$t_1(R)$ (6.82)$_1$	(Davis and Hill 1982) $t_2(R)$ (6.82)$_2$	(Hill and Kucera 1983a) $t_3(R)$ (6.82)$_3$	Numerical enthalpy scheme (Voller and Cross 1981)
0.1	0.8	0.0059	0.0059	0.0059	0.0059
	0.6	0.0220	0.0219	0.0218	0.0218
	0.4	0.0459	0.0449	0.0447	0.0445
	0.2	0.0750	0.0720	0.0709	0.0696
	0.0	0.1069	0.0997	0.0962	0.0875
1.0	0.8	0.0233	0.0232	0.0232	0.0232
	0.6	0.0820	0.0818	0.0817	0.0817
	0.4	0.1596	0.1585	0.1581	0.1578
	0.2	0.2395	0.2362	0.2345	0.2319
	0.0	0.3052	0.2970	0.2919	0.2752
10.0	0.8	0.1799	0.1799	0.1799	0.1798
	0.6	0.6129	0.6128	0.6128	0.6127
	0.4	1.1388	1.1385	1.1384	1.1380
	0.2	1.5976	1.5968	1.5963	1.5942
	0.0	1.8292	1.8272	1.8256	1.8087

approximations

$$t_1(R) = \frac{1}{\gamma}\left\{\frac{(1-R)^2}{2} - \frac{(1-R)^3}{(\gamma+3)}\right\},$$

$$t_2(R) = \frac{1}{\gamma}\left\{\frac{(1-R)^2}{2} - \frac{(1-R)^3}{(\gamma+3)} + \frac{\lambda_1(1-R)^4}{4(\gamma+3)^2}\right\}, \qquad (6.82)$$

$$t_3(R) = \frac{1}{\gamma}\left\{\frac{(1-R)^2}{2} - \frac{(1-R)^3}{(\gamma+3)} + \frac{\lambda_1(1-R)^4}{4(\gamma+3)^2} + \frac{\lambda_2(1-R)^5}{5(\gamma+3)^3}\right\},$$

are given in Table 6.3 for three values of α and five equally spaced boundary positions. Clearly, for both large and small α the successive approximations converge to the numerical values predicted by the enthalpy scheme. Moreover, even just the first two terms of (6.71) provide a reasonable approximation to the boundary motion. Figure 6.1 shows the temperature profiles resulting from taking one, two and three terms of (6.4) for $\alpha = 1.0$ and, clearly, the convergence is satisfactory. Results over a range of α indicate that the convergence is equally satisfactory for both large and small α and, moreover, that the temperature profiles have the same qualitative features as those from the numerical enthalpy scheme of Voller and Cross (1981), noted briefly in Section 1.6.

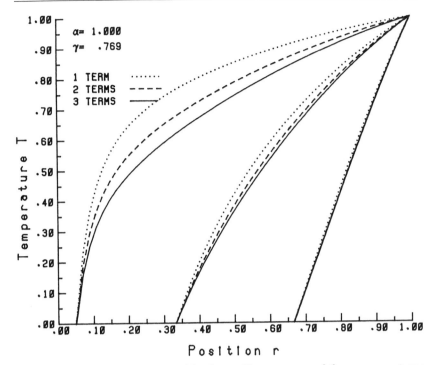

Figure 6.1 Temperature profiles resulting from taking one, two and three non-zero terms of (6.4) for spherical solidification with $\alpha = 1.0$ and β zero

6.3.2 Solution for surface heat loss ($\beta \neq 0$).

For β non-zero we find from (6.20) and (6.55) that $A_0(\rho)$ is identically zero, and therefore from (6.15) we see that the remaining $A_n(\rho)$ can be found from two integrations and the boundary conditions (6.56) and (6.57). On introducing two constants δ_1 and δ_2 defined by

$$\delta_1 = \alpha(\beta - 1), \qquad \delta_2 = 2 - \alpha, \tag{6.83}$$

we find that the first five functions $A_n(\rho)$ are as follows:

$$A_0(\rho) = 0, \qquad A_1(\rho) = \frac{1 - \rho}{\beta}, \qquad A_2(\rho) = \frac{\rho - 1}{2\alpha\beta^2}(\rho + 1 - 2\delta_1),$$

$$A_3(\rho) = \frac{1 - \rho}{6\alpha^2\beta^3}\left\{ \delta_1\rho^2 + (3\delta_2 - 11\delta_1)\rho + (3\delta_2 - 14\delta_1 + 6\delta_1^2) \right\},$$

$$\begin{aligned} A_4(\rho) = \frac{\rho - 1}{24\alpha^3\beta^4}\Big\{ &(4\delta_1 - \delta_2)\rho^3 + (4\delta_1 - \delta_2)(1 - 4\delta_1)\rho^2 \\ &+ (104\delta_1^2 - 56\delta_1\delta_2 - 68\delta_1 + 5\delta_2 + 12\delta_2^2 + 12)\rho \\ &- (24\delta_1^3 - 160\delta_1^2 + 68\delta_1\delta_2 + 68\delta_1 - 5\delta_2 - 12\delta_2^2 - 12) \Big\}. \end{aligned} \tag{6.84}$$

In particular, the values of $A_n'(1)$ needed to determine the motion of the boundary are

$$A_0'(1) = 0, \qquad A_1'(1) = -\frac{1}{\beta}, \qquad A_2'(1) = \frac{1 - \delta_1}{\alpha\beta^2},$$

$$A_3'(1) = -\frac{(\delta_1^2 - 4\delta_1 + \delta_2)}{\alpha^2\beta^3}, \tag{6.85}$$

$$A_4'(1) = \frac{1}{3\alpha^3\beta^4}(\delta_2 + 3\delta_2^2 + 3 - 15\delta_1\delta_2 - 3\delta_1^3 + 31\delta_1^2 - 16\delta_1).$$

From (2.82) and the above results we can show in the usual way that the motion of the boundary is given approximately by

$$t = \alpha\beta(1 - R) + [(\alpha + 1) - 2\alpha\beta]\frac{(1 - R)^2}{2}$$

$$+ [(\alpha\beta)^2 - (\alpha - 1)\alpha\beta - (\alpha + 1)]\frac{(1 - R)^3}{3\alpha\beta}$$

$$+ [(\alpha\beta)^2 - 2(\alpha + 2)\alpha\beta + (\alpha + 1)(\alpha + 2)]\frac{(1 - R)^4}{3(\alpha\beta)^2} + \dots, \tag{6.86}$$

so that in particular the estimate of the time t_c to complete solidification is

$$t_c = \frac{\alpha}{6}(1 + 2\beta) + \frac{1}{6(\alpha\beta)^2}\left\{7(\alpha\beta)^2 - 2(3\alpha + 5)\alpha\beta + 2(\alpha + 1)(\alpha + 2)\right\}. \tag{6.87}$$

We observe that t_c exceeds the lower bound (3.62) since

$$7(\alpha\beta)^2 - 2(3\alpha + 5)\alpha\beta + 2(\alpha + 1)(\alpha + 2) = 7\left(\alpha\beta - \frac{3\alpha + 5}{7}\right)^2$$

$$+ \tfrac{1}{7}(5\alpha^2 + 12\alpha + 3)$$

$$\geq 0.$$

Numerical results indicate that the above analysis applies only for sufficiently large β. For example, the condition

$$\beta \geq 1 + \alpha^{-1}, \tag{6.88}$$

is sufficient to ensure that the estimate (6.87) of t_c is less than the upper bound in (3.62). This follows since the required inequality is equivalent to

$$(\beta - 3)(\alpha\beta)^2 + (3\alpha + 5)\alpha\beta - (\alpha + 1)(\alpha + 2)$$
$$\geq (\alpha + 1)^2(\alpha^{-1} - 2) + (\alpha + 1)(2\alpha + 3)$$
$$= \frac{(\alpha + 1)(2\alpha + 1)}{\alpha}$$
$$\geq 0.$$

We note that on retaining only the terms of order one in the expressions (6.84) we find that (6.4) becomes

$$u(\rho, \tau) = \frac{(1 - \rho)\tau}{\beta}[1 - \phi + \phi^2 - \phi^3 + \ldots] + O\left(\frac{1}{\alpha}\right), \qquad (6.89)$$

where ϕ is defined by

$$\phi = \frac{(1 - \beta)\tau}{\beta}. \qquad (6.90)$$

On assuming the series in (6.89) is geometric and that $|\phi| < 1$, we may show in a straightforward manner that (6.89) gives rise to the pseudo-steady-state estimate (1.70). Similarly, on retaining only terms of order α and one in (6.86) we have

$$t = \alpha\left\{\beta(1 - R) + \frac{1 - 2\beta}{2}(1 - R)^2 + \frac{\beta - 1}{3}(1 - R)^3\right\}$$
$$+ \left\{\frac{(1 - R)^2}{2} + \frac{(\beta - 1)(1 - R)^3}{3\beta}[1 - \phi + \phi^2 + \ldots]\right\}$$
$$+ O\left(\frac{1}{\alpha}\right), \qquad (6.91)$$

and again the order one corrected motion (2.51) emerges on assuming the series in (6.91) is geometric with sum $(1 + \phi)^{-1}$. Figure 6.2 shows the temperature profiles for β non-zero resulting from taking three, four and five non-zero terms of (6.4) for $\alpha = 2.0$ and $\beta = 1.0$, where $A_5(\rho)$ is given in Hill and Kucera (1983b). Results for a range of α and β show that the convergence of these profiles varies considerably with α and β but for large α the convergence is generally better than that indicated in Fig. 6.2.

6.4   Freezing a liquid in a long circular cylindrical container

In this section we consider solutions of the form (6.4) to the problem of cylindrical solidification and, as in the previous section, the cases β zero

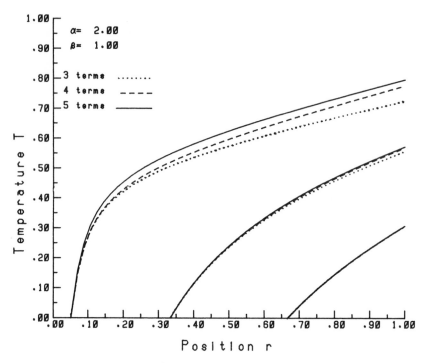

Figure 6.2 Temperature profiles resulting from taking three, four and five non-zero terms of (6.4) for spherical solidification with $\alpha = 2.0$ and $\beta = 1.0$

and non-zero must be treated separately. The following is due to Hill and Dewynne (1986). In terms of the variables (6.3) the problem reduces to (2.95)–(2.97) and on substituting (6.4) in (2.95) we again obtain (6.14) and (6.15), with γ given by (6.18) and the functions $f_n(\rho)$ defined by

$$f_1(\rho) = \left(\frac{A_1'(1)}{\alpha} - 2\gamma(\rho - 1)\right)\rho A_0',$$

$$f_2(\rho) = \frac{A_2'(1)}{\alpha} \rho A_0' + \left(\frac{A_1'(1)}{\alpha} - 2\gamma(\rho - 1)\right)(\rho A_1' - A_1) \qquad (6.92)$$

$$+ 2\left(\frac{A_1'(1)}{\alpha} - \gamma(\rho - 1)\right)\rho(\rho - 1)A_0',$$

and so on for $n \geq 3$. From (2.96) and (6.4) we find that the boundary conditions for β zero are exactly (6.53) and (6.54), while if β is non-zero we have (6.55) and

$$\beta A_1'(0) + A_0(0) = 1, \qquad A_1(1) = 0, \qquad (6.93)$$

$$\beta A_n'(0) + A_{n-1}(0) = 0, \qquad A_n(1) = 0 \qquad (n \geq 2). \qquad (6.94)$$

6.4.1 Solution for prescribed surface temperature $(\beta = 0)$.

For β zero $A_0(\rho)$ is again given by (6.39) with γ determined as usual from (6.40). Equations (6.58)–(6.65) apply here with $f_n(\rho)$ defined by (6.92). After a long calculation similar to that given in Appendix 2 we find

$$A_0'(1) = -\alpha\gamma, \qquad A_1'(1) = \frac{-\alpha\gamma^2}{2(\gamma + 3)},$$

$$A_2'(1) = \alpha\gamma\left(\frac{v_1}{(\gamma + 3)} - \frac{v_2(\gamma + 4)^2}{4(\gamma + 3)^2} - \frac{v_3}{6\gamma}\right), \tag{6.95}$$

where v_1, v_2 and v_3 are defined by

$$v_1 = \left\{\frac{(\gamma^4 + 4\gamma^3 - 5\gamma - 6) + \alpha\gamma(\gamma^3 + 3\gamma^2 - \gamma + 6)}{(\gamma^2 + 6\gamma + 3) + \alpha\gamma(\gamma + 5)}\right\},$$

$$v_2 = \left\{\frac{(2\gamma^3 + 3\gamma^2 - 3) + \alpha\gamma(2\gamma^2 + \gamma + 3)}{(\gamma^2 + 6\gamma + 3) + \alpha\gamma(\gamma + 5)}\right\}, \tag{6.96}$$

$$v_3 = \left\{\frac{(3\gamma^4 - 4\gamma^3 - 12\gamma^2 - 3) + \alpha\gamma(3\gamma^3 - 7\gamma^2 + \gamma + 3)}{(\gamma^2 + 6\gamma + 3) + \alpha\gamma(\gamma + 5)}\right\}.$$

Further, we have

$$A_1(\rho) = \frac{\alpha\gamma}{2}\rho\left\{(1 - \rho)\, e^{\gamma(1-\rho^2)/2} - \int_\rho^1 e^{\gamma(1-\xi^2)/2}\, d\xi\right.$$

$$\left. + \frac{1}{(\gamma + 3)}[e^{\gamma(1-\rho^2)/2} - 1]\right\}, \tag{6.97}$$

and we note that the expression for $A_2(\rho)$ is extremely lengthy and too complicated to give here.

From (2.97) and (6.4) we obtain the following approximation to the boundary motion,

$$t \approx \frac{1}{4\gamma}[(1 - R^2) + 2R^2 \log R]$$

$$+ \frac{1}{8(\gamma + 3)}[(1 - R^2) + 2R^2 \log R - 2R^2(\log R)^2]$$

$$+ \frac{1}{8\gamma}\left[\frac{A_2'(1)}{\alpha\gamma} + \frac{\gamma^2}{4(\gamma + 3)^2}\right]$$

$$\times \left[3(1 - R^2) + 6R^2 \log R - 6R^2(\log R)^2 + 4R^2(\log R)^3\right], \tag{6.98}$$

so that the time t_c for complete solidification is given approximately by

$$t_c \approx \frac{1}{4\gamma} + \frac{1}{8(\gamma + 3)} + \frac{3\gamma}{32(\gamma + 3)^2} + \frac{3A_2'(1)}{8\alpha\gamma^2}. \tag{6.99}$$

Using the first three terms of (1.57) we may show from $(6.95)_3$ and (6.96) that for large α we have

$$\frac{A_2'(1)}{\alpha\gamma} = \frac{\gamma}{15} + O(\gamma^2), \tag{6.100}$$

and (6.98) and (6.99) become, respectively,

$$t = \frac{\alpha}{4}[(1 - R^2) + 2R^2 \log R]$$

$$+ \frac{3}{30}[(1 - R^2) + 2R^2 \log R - 2R^2(\log R)^2]$$

$$+ \frac{1}{30}R^2(\log R)^3 + O\left(\frac{1}{\alpha}\right), \tag{6.101}$$

$$t_c = \frac{\alpha}{4} + \frac{3}{20} + O\left(\frac{1}{\alpha}\right). \tag{6.102}$$

As noted previously, (6.102) is remarkably close to the empirical result (6.7) of Voller and Cross (1981). Numerical values of (6.98) and (6.101) are given in Table 6.4 and compared with results from (4.94), the numerical enthalpy method of Voller and Cross (1981) and the finite difference scheme of Tao (1967). Clearly, (6.98) provides an accurate description of the boundary motion. Numerical results also confirm that (6.98) lies between the improved upper and lower bounds (3.72) and (3.91), respectively. This is shown in Fig. 6.3 for three values of α.

Table 6.4 Numerical values of the cylindrical boundary motions (6.98) and (6.101) for $\alpha = 1$ and β zero compared with (4.94) and the numerical schemes of Voller and Cross (1981) and Tao (1967)

$R(t)$	Integral iteration (4.94)	Voller and Cross (1981) (enthalpy)	Cylindrical boundary motion (6.98)	Tao (1967) (finite difference)	Large α approximation (6.101)
0.9	0.006	0.006	0.006	0.006	0.007
0.8	0.025	0.025	0.025	0.025	0.025
0.7	0.053	0.054	0.054	0.054	0.055
0.6	0.092	0.092	0.092	0.093	0.095
0.5	0.137	0.138	0.138	0.139	0.143
0.4	0.189	0.189	0.190	0.191	0.197
0.3	0.243	0.244	0.246	0.246	0.255
0.2	0.296	0.299	0.303	0.302	0.313
0.1	0.340	0.349	0.355	0.352	0.366
0.0	0.344	0.381	0.388	0.387	0.400

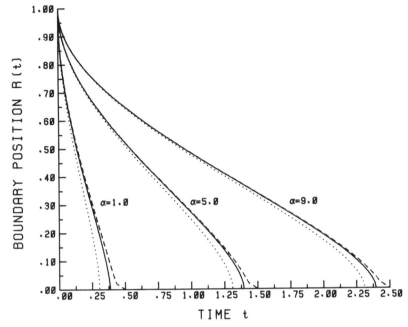

Figure 6.3 Cylindrical boundary motion for β zero from (6.98) compared with improved upper and lower bounds (3.72) and (3.91), respectively, for three values of α (from Hill and Dewynne (1986))

6.4.2 Solution for surface heat loss ($\beta \neq 0$). For β non-zero $A_0(\rho)$ and γ are again identically zero so that the $A_n(\rho)$ $(n \geq 1)$ are found by two integrations and using the boundary conditions (6.93) and (6.94). The final expressions are as follows:

$$A_0(\rho) = 0, \qquad A_1(\rho) = \frac{\rho - 1}{\beta}, \qquad A_2(\rho) = \frac{\rho - 1}{2\alpha\beta^2}(\rho + 2\alpha + 1),$$

$$A_3(\rho) = \frac{\rho - 1}{6\alpha^2\beta^3}\Big\{\alpha(2\beta - 1)\rho^2 + 2(4\alpha + 3 - 2\alpha\beta)\rho$$
$$+ (6\alpha^2 + 11\alpha + 6 - 4\alpha\beta)\Big\},$$

$$A_4(\rho) = \frac{\rho - 1}{24\alpha^3\beta^4}\Big\{(4\alpha^2\beta^2 - 4\alpha^2\beta + 2\alpha\beta - 3\alpha - 2)\rho^3$$
$$+ (-12\alpha^2\beta^2 + 28\alpha^2\beta + 18\alpha\beta - 12\alpha^2 - 11\alpha - 2)\rho^2$$
$$+ (12\alpha^2\beta^2 - 44\alpha^2\beta - 66\alpha\beta + 60\alpha^2 + 127\alpha + 70)\rho$$
$$+ (12\alpha^2\beta^2 - 60\alpha^2\beta - 66\alpha\beta + 24\alpha^3 + 104\alpha^2 + 151\alpha + 70)\Big\}.$$

$$(6.103)$$

In particular the values of $A'_n(1)$ needed to determine the motion of the boundary are:

$$A'_0(1) = 0, \qquad A'_1(1) = \frac{1}{\beta}, \qquad A'_2(1) = \frac{\alpha + 1}{\alpha\beta^2},$$

$$A'_3(1) = \frac{1}{\alpha^2\beta^3}\left\{(\alpha + 1)(\alpha + 2) - \alpha\beta\right\},$$

$$A'_4(1) = \frac{1}{3\alpha^3\beta^4}\left\{(\alpha + 1)(3\alpha^2 + 16\alpha + 17) + 2\alpha^2\beta^2 - 10\alpha^2\beta - 14\alpha\beta\right\}.$$

$$(6.104)$$

From these results and (2.97) we may deduce the following approximating expression for the motion of the boundary:

$$t \approx \frac{\alpha\beta}{2}(1 - R^2) + \frac{\alpha + 1}{4}\left\{(1 - R^2) + 2R^2\log R\right\}$$

$$+ \frac{\alpha\beta - (\alpha + 1)}{4\alpha\beta}\left\{(1 - R^2) + 2R^2\log R - 2R^2(\log R)^2\right\}$$

$$+ \frac{1}{12\alpha^2\beta^2}\left\{(\alpha\beta)^2 - 2\alpha\beta(\alpha + 2) + 2(\alpha + 1)(\alpha + 2)\right\}$$

$$\times \left\{3(1 - R^2) + 6R^2\log R - 6R^2(\log R)^2 + 4R^2(\log R)^3\right\}, \qquad (6.105)$$

so that for the time to complete solidification we have

$$t_c = \frac{\alpha}{4}(1 + 2\beta) + \frac{1}{4(\alpha\beta)^2}\left\{3(\alpha\beta)^2 - (3\alpha + 5)\alpha\beta + 2(\alpha + 1)(\alpha + 2)\right\},$$

$$(6.106)$$

and this estimate always exceeds the pseudo-steady-state lower bound (3.64) since

$$[3(\alpha\beta)^2 - (3\alpha + 5)\alpha\beta + 2(\alpha + 1)(\alpha + 2)]$$

$$= 3\left(\alpha\beta - \frac{3\alpha + 5}{6}\right)^2 + \tfrac{1}{12}(15\alpha^2 + 42\alpha + 23)$$

$$\geq 0.$$

Again, numerical results indicate that the above formulae are meaningful either for large α or for sufficiently large β and, for example, the condition

$$\beta \geq 1 + (2\alpha)^{-1}, \qquad (6.107)$$

is sufficient to ensure that (6.106) is less than the upper bound (3.64) since

$$2(\beta - 1)(\alpha\beta)^2 + (3\alpha + 5)\alpha\beta - 2(\alpha + 1)(\alpha + 2)$$

$$\geq \frac{(2\alpha + 1)^2}{4\alpha} + \frac{2\alpha + 1}{2}(3\alpha + 5) - 2(\alpha + 1)(\alpha + 2)$$

$$= \frac{1}{4\alpha}\left\{4\alpha^3 + 6(\alpha - \tfrac{1}{6})^2 + \tfrac{5}{6}\right\}$$

$$\geq 0.$$

Again we note that the order one terms from (6.103) give

$$u(\rho, \tau) = (\rho - 1)\frac{\tau}{\beta}\left\{1 + \left(\frac{\tau}{\beta}\right) + \left(\frac{\tau}{\beta}\right)^2 + \ldots\right\} + O\left(\frac{1}{\alpha}\right), \qquad (6.108)$$

which formally sums to the pseudo-steady-state result (1.73) assuming the series is geometric and that $|\tau| < \beta$. However, although the order α term of (6.105) gives the pseudo-steady-state motion (1.74), it is not entirely clear how the order one terms sum to the order one corrected motion (2.61). Figure 6.4 shows the cylindrical boundary motion for β non-zero

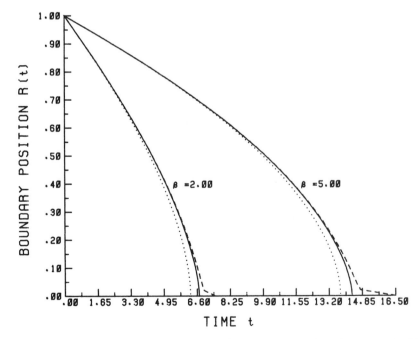

Figure 6.4 Cylindrical boundary motion for β non-zero from (6.105) compared with improved upper and lower bounds (3.72) and (3.91), respectively, for $\alpha = 5.0$ and two values of β (from Hill and Dewynne (1986))

contained by the improved upper and lower bounds (3.72) and (3.91), respectively, for $\alpha = 5.0$ and two values of β.

6.5 Alternative solution for spherical freezing with no surface cooling

In this section we again examine the problem of spherical solidification with β zero by means of a series solution of the form (6.4) but with variables ρ, τ and u defined by (6.13). The analysis follows Hill and Kucera (1984). In terms of the variables (6.13), the problem (2.4)–(2.6) with β zero becomes

$$\alpha u_{\rho\rho} = (1 + \tau)^4 (1 + \rho\tau)^{-4} u_\rho (1, \tau)(\rho u_\rho - \tau u_\tau), \tag{6.109}$$

$$u(0, \tau) = 1, \qquad u(1, \tau) = 0, \tag{6.110}$$

$$u_\rho(1, \tau) = -\frac{\alpha\tau}{(1 + \tau)^4} \frac{d\tau}{dt}, \qquad \tau(0) = 0. \tag{6.111}$$

On substituting (6.4) into (6.109) we again have (6.14) and (6.15) where

$$f_1(\rho) = \left(\frac{A_1'(1)}{\alpha} + 4\gamma(\rho - 1) \right) \rho A_0',$$

$$f_2(\rho) = \frac{A_2'(1)}{\alpha} \rho A_0' + \left(\frac{A_1'(1)}{\alpha} + 4\gamma(\rho - 1) \right)(\rho A_1' - A_1) \tag{6.112}$$

$$- 2\left(\frac{2A_1'(1)}{\alpha} + \gamma(5\rho - 3) \right) \rho(\rho - 1)A_0',$$

while (6.4) and (6.110) give precisely the boundary conditions (6.53) and (6.54). Thus, $A_0(\rho)$ and γ are as usual given by (1.49) and (1.43), respectively. Following Section 6.3.1 and Appendix 2 we may deduce in a completely analogous manner

$$A_0'(1) = -\alpha\gamma, \qquad A_1'(1) = \frac{\alpha\gamma^2}{(\gamma + 3)},$$

$$A_2'(1) = \alpha\gamma \left[\frac{2\beta_1}{(\gamma + 3)} - \left(\frac{\gamma + 4}{\gamma + 3} \right)^2 \beta_2 \right], \tag{6.113}$$

where β_1 and β_2 are defined by

$$\beta_1 = \left\{ \frac{(\gamma^4 + 6\gamma^3 + 6\gamma^2 - 9) + \alpha\gamma(\gamma^3 + 5\gamma^2 + 3\gamma + 9)}{(\gamma^2 + 6\gamma + 3) + \alpha\gamma(\gamma + 5)} \right\},$$

$$\beta_2 = \left\{ \frac{(2\gamma^3 + 3\gamma^2 - 3) + \alpha\gamma(2\gamma^2 + \gamma + 3)}{(\gamma^2 + 6\gamma + 3) + \alpha\gamma(\gamma + 5)} \right\}. \tag{6.114}$$

Further with γ_1 and γ_2 defined by

$$\gamma_1 = \frac{A_1'(1)}{\alpha\gamma}, \qquad \gamma_2 = \frac{A_2'(1)}{\alpha\gamma}, \qquad (6.115)$$

we find from (6.111) that the motion of the boundary is obtained by integrating

$$dt = \frac{\tau \, d\tau}{\gamma(1+\tau)^4(1-\gamma_1\tau - \gamma_2\tau^2)}. \qquad (6.116)$$

On using $(6.13)_2$ we obtain

$$t = \frac{1}{\gamma}\int_R^1 (1-\xi)\psi(\xi) \, d\xi, \qquad (6.117)$$

where the function $\psi(\xi)$ is defined by

$$\psi(\xi) = \xi^3[(1+\gamma_1-\gamma_2)\xi^2 - (\gamma_1-2\gamma_2)\xi - \gamma_2]^{-1}. \qquad (6.118)$$

From the Taylor series about $\xi = 1$, namely

$$\psi(\xi) = \psi(1) + (\xi - 1)\frac{d\psi}{d\xi}(1) + \frac{(\xi-1)^2}{2}\frac{d^2\psi}{d\xi^2}(1) + \ldots, \qquad (6.119)$$

and

$$\psi(1) = 1, \qquad \frac{d\psi}{d\xi}(1) = 1 - \gamma_1, \qquad \frac{d^2\psi}{d\xi^2}(1) = 2(\gamma_2 + \gamma_1^2), \qquad (6.120)$$

we find that (6.117) yields

$$t = \frac{1}{\gamma}\left\{\frac{(1-R)^2}{2} - \frac{1-\gamma_1}{3}(1-R)^3 + \frac{\gamma_2 + \gamma_1^2}{4}(1-R)^4 + \ldots\right\}. \qquad (6.121)$$

On noting the identity

$$\gamma_2 + \gamma_1^2 = \frac{\mu_1 - 3\gamma}{(\gamma+3)^2}, \qquad (6.122)$$

we may readily confirm that (6.121) coincides exactly with the first three terms of (6.71).

Alternatively, we may deduce expressions for $A_1(\rho)$ and $A_2(\rho)$ and hence (6.115) more simply as follows. With the convention that bars denote variables and functions defined by (6.2) and used in Section 6.3.1 we have

$$\bar{\rho} = \frac{1-r}{1-R} = \frac{\rho(1+\tau)}{1+\rho\tau} = \rho + \rho(1-\rho)\tau - \rho^2(1-\rho)\tau^2 + \ldots,$$

$$\bar{\tau} = 1 - R = -\frac{\tau}{1+\tau} = -\tau + \tau^2 + \ldots. \qquad (6.123)$$

On equating expressions for the temperature $T(r, t)$ we have

$$A_0(\rho) + A_1(\rho)\tau + A_2(\rho)\tau^2 + \ldots$$
$$= (1 + \rho\tau)[\bar{A}_0(\bar{\rho}) + \bar{A}_1(\bar{\rho})\bar{\tau} + \bar{A}_2(\bar{\rho})\bar{\tau}^2 + \ldots], \tag{6.124}$$

and on further equating coefficients of τ^n ($n = 0, 1, 2, \ldots$) we obtain

$$A_1(\rho) = \bar{A}_1(\rho) + \rho(1 - \rho)A_0'(\rho) + \rho A_0(\rho),$$

$$A_2(\rho) = \bar{A}_2(\rho) + (1 - \rho)[\rho\bar{A}_1'(\rho) - \bar{A}_1(\rho)] + \frac{\rho^2(1 - \rho)^2 A_0''(\rho)}{2},$$

$$\tag{6.125}$$

where $\bar{A}_1(\rho)$ and $\bar{A}_2(\rho)$ are defined explicitly by $(6.66)_1$ and $(6.66)_2$, respectively. From (6.66) we obtain

$$A_1(\rho) = -\alpha\gamma\rho\left\{(1 - \rho)\, e^{\gamma(1-\rho^2)/2} - \int_\rho^1 e^{\gamma(1-\xi^2)/2}\, d\xi \right.$$

$$\left. + \frac{1}{(\gamma + 3)}[e^{\gamma(1-\rho^2)/2} - 1]\right\},$$

$$A_2(\rho) = \bar{A}_2(\rho) + \alpha\gamma^2\rho^3(1 - \rho)\left\{\frac{1}{(\gamma + 3)} + \frac{1 - \rho}{2}\right\} e^{\gamma(1-\rho^2)/2}, \tag{6.126}$$

with $\bar{A}_2(\rho)$ given by $(6.66)_2$. We note that the identity (6.122) follows immediately from $(6.126)_2$ since

$$A_2'(1) = \bar{A}_2'(1) - \frac{\alpha\gamma^2}{(\gamma + 3)}, \tag{6.127}$$

which using $(6.68)_3$, $(6.69)_1$ and (6.115) can be shown to be equivalent to (6.122).

Thus, although the boundary motion is in agreement with that obtained previously in Section 6.3.1, numerical results indicate that the two temperature profiles from (6.4) agree only for small times. For small α and times close to complete solidification the differences in the two temperature values become quite pronounced due to the singular nature of the expansion variable τ given by (6.13).

6.6 Additional symbols used

a, c	parameters in $_1F_1(a, c; y)$, (6.60)
$(a)_k$	symbol defined by (6.61)
$A(y)$	solution of (6.58)
$A_n(\rho)$	coefficients in series solution (6.4)

$A_{n1}(\rho)$, $A_{n2}(\rho)$	linearly independent solutions of (6.58) defined by (6.59) and with Wronskian given by (6.43)
$B_n(\rho)$	coefficients in series solution (6.5)
$B_{n1}(\rho)$, $B_{n2}(\rho)$	linearly independent solutions of homogeneous equation in (6.35) defined by (6.42) and with Wronskian given by (6.43)
C	arbitrary constant in (6.11)
C_1, C_2	arbitrary constants in (6.20)
C_{n1}, C_{n2}	arbitrary constants in (6.62)
$f(r)$	arbitrary function in general boundary-fixing transformation (6.8)
$f_n(\rho)$	right-hand side of (6.15)
$_1F_1(a, c; y)$	confluent hypergeometric function, (6.60)
$g_n(\rho)$	right-hand side of (6.35)
I_n	integrals defined by (6.75)
K_λ	function defined by (3.95)
$t_1(R)$, $t_2(R)$, $t_3(R)$	successive approximations to spherical boundary motion, (6.82)
$v(\rho, \tau)$	temperature for planar problem, (6.29)
y	variable $-\gamma\rho^2/2$, (6.58)

Greek symbols

β_1, β_2	constants defined by (6.114)
γ_1, γ_2	constants defined by (6.115)
δ_1, δ_2	constants defined by (6.83)
λ_1, λ_2	constants defined by (6.72)
μ	constant defined by (6.67)
μ_1, μ_2	constants defined by (6.69)
ν_1, ν_2, ν_3	constants defined by (6.96)
σ	large-time expansion variable for the planar problem, $(6.29)_2$
ϕ	variable defined by (6.90)
$\psi(\xi)$	function defined by (6.118)
$\omega(\rho)$	Wronskian defined by (6.43)

Conventions

Primes throughout the chapter denote differentiation with respect to ρ. ρ and τ have different meanings in different sections. In Sections 6.2, 6.3, 6.4 and 6.5 ρ and τ are defined, respectively, by (6.1), (6.2), (6.3) and (6.13). In Section 6.5 bars on variables and functions denote quantities used and derived in Section 6.3.1.

7

Two simultaneous chemical reactions

7.1 Introduction

In this chapter we extend some of the results of previous chapters to a slightly more complicated problem. We consider the 'shrinking-core' model for fluid–solid reactions proposed by Krishnamurthy and Shah (1979), which arises from an essentially instantaneous fluid–solid reaction (giving rise to a moving reaction front) together with a slower first-order reaction occurring in the region behind the reaction front. This problem arises in the oxydesulphurization of coal which contains both organic and inorganic sulphur. The oxidation of the former and carbon is very slow, while that of the latter is very rapid. We consider a sphere of radius a consisting of an inert solid matrix, in which various solid reactants are supported, and with a porous structure, saturated with fluid, which allows a chemical species in the fluid to diffuse into the inert matrix. At time zero the surface $r^* = a$ is subjected to a concentration T_0 of solute, and is held at T_0 thereafter. The solute is assumed to react instantaneously with one of the solid reactants, giving rise to a reaction front $R^*(t^*)$, while in the region between the surface and the reaction front a second slower reaction occurs. We assume this second slow reaction is first order with respect to the chemical reactant in the fluid. In non-dimensional variables we may state the problem as

$$\frac{\partial T}{\partial t} = \frac{\partial^2 T}{\partial r^2} + \frac{2}{r}\frac{\partial T}{\partial r} - k^2 T, \qquad R(t) < r < 1, \tag{7.1}$$

$$T(1, t) + \beta \frac{\partial T}{\partial r}(1, t) = 1, \qquad T(R(t), t) = 0, \tag{7.2}$$

$$\frac{\partial T}{\partial r}(R(t), t) = -\alpha \frac{dR}{dt}, \qquad R(0) = 1, \tag{7.3}$$

where $T(r, t)$ denotes the non-dimensional concentration and $R(t)$ the non-dimensional location of the reaction front at time t. The non-dimensional variables and constants α, β and k are defined in terms of the 'starred' physical quantities by

$$r = \frac{r^*}{a}, \qquad R = \frac{R^*}{a}, \qquad t = \frac{Dt^*}{a^2}, \qquad T = \frac{T^*}{T_0}, \tag{7.4}$$

$$\alpha = \frac{\rho_1 \omega}{T_0}, \qquad k^2 = \frac{a^2 k_1}{D}, \qquad \beta = \frac{1}{ah_D} \quad \text{or} \quad 0, \tag{7.5}$$

where D is the diffusivity of the chemical reactant through the fluid, ρ_1 is the density of the solid, k_1 is the first-order rate constant for the slow reaction, ω is the stoichiometric coefficient for the rapid reaction and h_D is the surface mass transfer coefficient. The value of β depends on whether there is mass transfer at the surface or not, and we assume k_1 and therefore k^2 to be positive quantities. Thus, the problem of two simultaneous chemical reactions (7.1)–(7.3) is the same as the problem of spherical solidification (2.4)–(2.6), except that (2.4) is replaced by (7.1). This problem has been noted previously together with its pseudo-steady-state solution and boundary motion (see Eqs. (1.63), (1.77) and (1.78)). For our purposes this problem is an appropriate extension of (2.4)–(2.6), sufficient to demonstrate how results of previous chapters might be extended.

In the following section we make use of the boundary-fixing transformation (7.12) to deduce two terms of the large Stefan number expansion for $T(r, t)$ which, together with the exact integral formula (7.53) for the boundary motion, enables three terms of the large Stefan number expansion for the boundary motion to be obtained. In Section 7.3 we deduce an integral formulation of the problem based on a direct integration of (7.7) which, following Krishnamurthy and Shah (1979), we may utilize as an integral iteration scheme. In Section 7.4 we deduce an alternative integral formulation by means of a Green's function. This formulation may also be used as the basis of an integral iteration scheme but more importantly permits the exact integral formula (7.53) for the boundary motion and enables the important inequality $T \leq T_{pss}$ to be established. In Section 7.5 we discuss various bounds for the motion of the boundary, including those resulting from $T \leq T_{pss}$ and the improved lower bound arising from a further integration of (7.53). In the final section of the chapter we note the appropriate generalizations to k non-zero of the formal series solution (4.18) and the integral equation (4.51).

We note here that in terms of the standard transformation

$$T(r, t) = \frac{v(r, t)}{r}, \tag{7.6}$$

the problem (7.1)–(7.3) becomes

$$\frac{\partial v}{\partial t} = \frac{\partial^2 v}{\partial r^2} - k^2 v, \qquad R(t) < r < 1, \tag{7.7}$$

$$(1 - \beta)v(1, t) + \beta \frac{\partial v}{\partial r}(1, t) = 1, \qquad v(R(t), t) = 0, \tag{7.8}$$

$$\frac{\partial v}{\partial r}(R(t), t) = -\alpha R \frac{dR}{dt}, \qquad R(0) = 1, \tag{7.9}$$

and that the usual transformation

$$v(r, t) = e^{-k^2 t} \bar{\bar{v}}_1(r, t), \tag{7.10}$$

or Danckwert's transformation (see Crank 1964, p. 125)

$$v(r, t) = e^{-k^2 t} \bar{\bar{v}}_1(r, t) + k^2 \int_0^t e^{-k^2 \eta} \bar{\bar{v}}_1(r, \eta) \, d\eta, \tag{7.11}$$

where $\bar{\bar{v}}_1(r, t)$ is a solution of the one-dimensional heat equation, appear not to be effective in the context of Stefan problems.

We also note that some of the results of this chapter are either generalizations of, or based on, results from Hill (1984a, 1984b) and Dewynne and Hill (1985). Moreover, throughout the chapter we restrict our attention to the spherical geometry, since fluid flow through a packed bed of chemically reacting spherical particles is a common industrial process. However, a similar partial differential equation arises in the linear flow of heat in rods (see, for example, Carslaw and Jaeger 1965, p. 133) and some of the results of this chapter for slab and cylindrical geometries may be found in Hill (1984b) and Dewynne and Hill (1985).

7.2 Large Stefan number expansion

In this section we use the boundary-fixing transformation

$$\rho = \frac{\sinh k(1 - r)}{\sinh k(1 - R)}, \qquad \tau = \sinh k(1 - R), \qquad v(r, t) = u(\rho, \tau), \tag{7.12}$$

to obtain the large α expansion of (7.7)–(7.9). Although the choice of the boundary-fixing transformation is to a certain extent arbitrary, we observe that the above variables are 'essentially' those defined by (2.79) for spherical solidification in the limit k tending to zero. This identification can be made precise simply by using τ/k in place of τ. However, for our purposes (7.12) is more convenient and in particular we have the relations

$$\sinh k(1 - r) = \rho\tau, \qquad \cosh k(1 - r) = (1 + \rho^2\tau^2)^{1/2},$$
$$\sinh k(1 - R) = \tau, \qquad \cosh k(1 - R) = (1 + \tau^2)^{1/2}. \qquad (7.13)$$

In terms of (7.12) the problem (7.7)–(7.9) becomes

$$(1 + \rho^2\tau^2)u_{\rho\rho} + \tau^2(\rho u_\rho - u) = \frac{1 + \tau^2}{\alpha R(t)} u_\rho(1, \tau)(\rho u_\rho - \tau u_\tau), \qquad (7.14)$$

$$\beta k u_\rho(0, \tau) = \tau[(1 - \beta)u(0, \tau) - 1], \qquad u(1, \tau) = 0, \qquad (7.15)$$

$$u_\rho(1, \tau) = -\frac{\alpha R(t)\tau}{k^2(1 + \tau^2)} \frac{d\tau}{dt}, \qquad \tau(0) = 0, \qquad (7.16)$$

where $R(t)$ in terms of τ is given explicitly by

$$R(t) = 1 - \frac{\sinh^{-1}(\tau)}{k} = 1 - \frac{\log[\tau + (1 + \tau^2)^{1/2}]}{k}. \qquad (7.17)$$

On looking for a solution for large α of the form

$$u(\rho, \tau) = u_0(\rho, \tau) + \frac{u_1(\rho, \tau)}{\alpha} + \frac{u_2(\rho, \tau)}{\alpha^2} + O\left(\frac{1}{\alpha^3}\right), \qquad (7.18)$$

we obtain the following boundary-value problems for the first two terms:

$$(1 + \rho^2\tau^2)u_{0\rho\rho} + \tau^2(\rho u_{0\rho} - u_0) = 0,$$
$$\beta k u_{0\rho}(0, \tau) = \tau[(1 - \beta)u_0(0, \tau) - 1], \qquad u_0(1, \tau) = 0, \qquad (7.19)$$

and

$$(1 + \rho^2\tau^2)u_{1\rho\rho} + \tau^2(\rho u_{1\rho} - u_1) = \frac{1 + \tau^2}{R(t)} u_{0\rho}(1, \tau)(\rho u_{0\rho} - \tau u_{0\tau}),$$
$$(7.20)$$
$$\beta k u_{1\rho}(0, \tau) = \tau(1 - \beta)u_1(0, \tau), \qquad u_1(1, \tau) = 0.$$

On noting that the general solution of (7.19)$_1$ takes the form

$$u_0(\rho, \tau) = A(\tau)\rho + B(\tau)(1 + \rho^2\tau^2)^{1/2}, \qquad (7.21)$$

where $A(\tau)$ and $B(\tau)$ denote arbitrary functions of τ, we may readily deduce, from the boundary conditions (7.19)$_2$, the following expression for $u_0(\rho, \tau)$:

$$u_0(\rho, \tau) = \left\{ \frac{\tau(1 + \rho^2\tau^2)^{1/2} - \rho\tau(1 + \tau^2)^{1/2}}{(1 - \beta)\tau + \beta k(1 + \tau^2)^{1/2}} \right\}. \qquad (7.22)$$

On using the relations (7.13), this expression can be shown to yield precisely the pseudo-steady-state approximation (1.77). From (7.22) we find that $(7.20)_1$ becomes

$$(1 + \rho^2\tau^2)u_{1\rho\rho} + \tau^2(\rho u_{1\rho} - u_1) = \frac{\tau^2[(1-\beta)\rho\tau + \beta k(1 + \rho^2\tau^2)^{1/2}]}{R(t)[(1-\beta)\tau + \beta k(1 + \tau^2)^{1/2}]^3},$$

$$(7.23)$$

and by variation of parameters we may show in the usual way that the solution $u_1(\rho, \tau)$ is given by

$$u_1(\rho, \tau) = \frac{1}{2R(t)[(1-\beta)\tau + \beta k(1 + \tau^2)^{1/2}]^4}\Big\{[(1-\beta)\tau + \beta k(1 + \tau^2)^{1/2}]$$
$$\times \{(1-\beta)[(1 + \rho^2\tau^2)^{1/2}\sinh^{-1}(\rho\tau) - \rho\tau] + \beta k \rho\tau \sinh^{-1}(\rho\tau)\}$$
$$- [(1-\beta)\rho\tau + \beta k(1 + \rho^2\tau^2)^{1/2}]$$
$$\times \{(1-\beta)[(1 + \tau^2)^{1/2}\sinh^{-1}(\tau) - \tau] + \beta k\tau \sinh^{-1}(\tau)\}\Big\}.$$

$$(7.24)$$

We observe that in the limit of k zero we recover the expression $(2.90)_2$ and that $u_1(\rho, \tau)$ is, as usual, singular as R tends to zero. On introducing the function $\phi(r)$ defined by

$$\phi(r) = (1 - \beta)\sinh k(1 - r) + \beta k \cosh k(1 - r),\qquad(7.25)$$

we may show altogether that we have

$$T(r, t) = \frac{\sinh k(r - R)}{r\phi(R)}$$
$$+ \frac{\left\{\begin{array}{c}(1 - R)\phi'(R)\phi(r) - (1 - r)\phi'(r)\phi(R)\\ + (1 - \beta)\beta k \sinh k(r - R)\end{array}\right\}}{2\alpha r R \phi(R)^4}$$
$$+ O\left(\frac{1}{\alpha^2}\right),\qquad(7.26)$$

where primes denote differentiation with respect to the argument indicated. Although we may evaluate the order one corrected motion via (7.18), (7.22), (7.26) and the Stefan condition (7.16) it is simpler to use the exact integral formula (7.53) which, moreover, gives an expression for the term of order α^{-1}.

From (7.53) and the above equations we find

$$t(R) = \alpha\tau_0(R) + \tau_1(R) + \frac{\tau_2(R)}{\alpha} + O\left(\frac{1}{\alpha^2}\right),\qquad(7.27)$$

where $\tau_0(R)$, $\tau_1(R)$ and $\tau_2(R)$ are given by

$$\tau_0(R) = -\frac{R\phi'(R) - \phi(R) + k}{k^3},$$

$$\tau_1(R) = -\frac{(1-R)\phi'(R) + (1-\beta)\sinh k(1-R)}{2k^2\phi(R)}, \qquad (7.28)$$

$$\tau_2(R) = \int_R^1 \phi(\xi) \frac{\left\{\begin{array}{l}(1-R)\phi'(R)\phi(\xi) - (1-\xi)\phi'(\xi)\phi(R) \\ \qquad + (1-\beta)\beta k \sinh k(\xi - R)\end{array}\right\}}{2kR\phi(R)^4} \, d\xi,$$

where $\alpha\tau_0(R)$ is simply the pseudo-steady-state motion $t_{pss}(R)$ given explicitly by (1.78). We note that the order one contribution to the boundary motion is given explicitly in (7.68) and that, although the integrals for $\tau_2(R)$ can all be evaluated in elementary terms (see integrals (7.64) and (7.67)), the final result does not simplify and is therefore not given.

The above equations simplify considerably for the case of prescribed concentration at the surface (β zero). The final results are as follows:

$$T(r, t) = \frac{\sinh k(r - R)}{r \sinh k(1 - R)}$$

$$+ \frac{k\left\{\begin{array}{l}(1-r)\cosh k(1-r)\sinh k(1-R) \\ \qquad - (1-R)\cosh k(1-R)\sinh k(1-r)\end{array}\right\}}{2\alpha r R \sinh^4 k(1 - R)}$$

$$+ O\!\left(\frac{1}{\alpha^2}\right), \qquad (7.29)$$

while for the boundary motion we have

$$\tau_0(R) = \frac{kR\cosh k(1 - R) + \sinh k(1 - R) - k}{k^3},$$

$$\tau_1(R) = \frac{k(1 - R)\coth k(1 - R) - 1}{2k^2}, \qquad (7.30)$$

$$\tau_2(R) = \frac{\left\{\begin{array}{l}\cosh k(1-R)[2k^2(1-R)^2 - \sinh^2 k(1-R)] \\ \qquad - k(1-R)\sinh k(1-R)\end{array}\right\}}{8k^2R\sinh^4 k(1 - R)},$$

and these results are consistent with (2.108), (2.109) and (2.111) in the limit of k tending to zero.

7.3 Integral formulation by integration

In this section, following Krishnamurthy and Shah (1979), we formulate an integral iteration scheme by integrating (7.7) directly. Using this method, Eq. (7.33) for the boundary motion does not permit a formal integration, in contrast to the Green's function approach used in the following section. On integrating (7.7) directly and using (7.9)$_1$ we find

$$\frac{\partial v}{\partial r} = \frac{\partial}{\partial t} \int_R^r [\alpha\xi + v(\xi, t)] \, d\xi + k^2 \int_R^r v(\xi, t) \, d\xi, \tag{7.31}$$

and from a further integration of (7.31) and (7.8)$_2$ we have

$$v(r, t) = \frac{\partial}{\partial t} \int_R^r (r - \xi)[\alpha\xi + v(\xi, t)] \, d\xi + k^2 \int_R^r (r - \xi)v(\xi, t) \, d\xi. \tag{7.32}$$

From these equations and the surface condition (7.8)$_1$ we obtain

$$\frac{d}{dt} \int_R^1 [1 + (\beta - 1)\xi][\alpha\xi + v(\xi, t)] \, d\xi$$

$$+ k^2 \int_R^1 [1 + (\beta - 1)\xi]v(\xi, t) \, d\xi = 1, \tag{7.33}$$

which appears not to admit a simple formal integration. However, if we define the integral $I(R)$ by

$$I(R) = \int_R^1 [1 + (\beta - 1)\xi]v^\dagger(\xi, R) \, d\xi, \tag{7.34}$$

with $v^\dagger(r, R)$ denoting $v(r, t)$ but with independent variables (r, R), then we see from (7.33) that the boundary motion is obtained by integrating

$$dt = -\frac{\alpha R[1 + (\beta - 1)R] - dI/dR}{1 - k^2 I(R)} \, dR, \tag{7.35}$$

so that we have

$$t = \frac{1}{k^2} \log \frac{1}{1 - k^2 I(R)} + \int_R^1 \frac{\alpha\xi[1 + (\beta - 1)\xi]}{1 - k^2 I(\xi)} \, d\xi. \tag{7.36}$$

Further, if for convenience of notation we introduce

$$J(r, R) = \int_R^r (r - \xi)v^\dagger(\xi, R) \, d\xi, \tag{7.37}$$

then from (7.32) and (7.33) we have

$$\frac{v^\dagger(r, R) - k^2 J(r, R)}{1 - k^2 I(R)} = \left(\frac{\alpha R(r - R) - \partial J/\partial R}{\alpha R[1 + (\beta - 1)R] - dI/dR} \right), \tag{7.38}$$

and the procedure of Krishnamurthy and Shah (1979) is to employ (7.36) and (7.38) as the basis for an iterative scheme.

Thus, with v_{pss} as the initial estimate and

$$I_{\mathrm{pss}}(R) = k^{-2}\left\{1 - \frac{k[1 + (\beta - 1)R]}{\phi(R)}\right\}, \tag{7.39}$$

$$J_{\mathrm{pss}}(r, R) = \frac{\sinh k(r - R) - k(r - R)}{k^2 \phi(R)}, \tag{7.40}$$

where $\phi(R)$ is defined by (7.25), we have from (7.36)

$$\bar{t}_1(R) = t_{\mathrm{pss}}(R) + \frac{1}{k^2}\log\left\{\frac{\phi(R)}{k[1 + (\beta - 1)R]}\right\}, \tag{7.41}$$

where $t_{\mathrm{pss}}(R)$ is given explicitly by (1.78), while from (7.38) we may simplify the result to obtain

$$\bar{v}_1(r, R) = v_{\mathrm{pss}}(r, R)$$
$$+ \frac{k\{[1 + (\beta - 1)R]\phi(r) - [1 + (\beta - 1)r]\phi(R)\}}{\phi(R)\left\{\begin{array}{l}\alpha kR[1 + (\beta - 1)R]\phi(R)^2 + (\beta - 1)\phi(R) \\ \qquad - [1 + (\beta - 1)R]\phi'(R)\end{array}\right\}}. \tag{7.42}$$

We note that the bars are used to distinguish these estimates with those obtained in the following section and that (7.41) and (7.42) are in agreement with the corresponding results of Krishnamurthy and Shah (1979) for the limiting case β zero (apart from a minor printing error in their formula for $\bar{t}_1(R)$). Further, we observe that (7.41) provides the following estimate for the time to complete reaction:

$$\bar{t}_1(0) = t_{\mathrm{pssc}} + \frac{1}{k^2}\log\left\{\frac{(1 - \beta)\sinh k + \beta k \cosh k}{k}\right\}, \tag{7.43}$$

where t_{pssc} is given by (1.79).

To obtain further iterations using (7.42) becomes more complicated. However, using (7.34), (7.36) and (7.42) we may give an expression $\bar{t}_2(R)$ for the boundary motion, involving an integral which would need to be evaluated numerically. The final result is as follows:

$$\bar{t}_2(R) = t_{\mathrm{pss}}(R) + \frac{1}{k^2}\log\frac{1}{1 - k^2 I_1(R)}$$
$$+ \alpha \int_R^1 \xi\Big\{3(\beta - 1)[1 + (\beta - 1)\xi]\phi(\xi) - 3[1 + (\beta - 1)\xi]^2\phi'(\xi)$$
$$- k^2(1 - R)[(1 + \beta + \beta^2) + (\beta - 1)(\beta + 2)R + (\beta - 1)^2 R^2]\phi(\xi)\Big\}$$
$$\times \{3k^2\Delta_1(\xi)\}^{-1}\,\mathrm{d}\xi, \tag{7.44}$$

where $[1 - k^2 I_1(R)]^{-1}$ and $\Delta_1(R)$ are given by

$$\frac{1}{1 - k^2 I_1(R)} = \frac{\left\{\begin{array}{c} \alpha k R[1 + (\beta - 1)R]\phi(R)^2 + (\beta - 1)\phi(R) \\ - [1 + (\beta - 1)R]\phi'(R) \end{array}\right\}}{k^2 \Delta_1(R)}, \qquad (7.45)$$

$$\Delta_1(R) = \alpha R[1 + (\beta - 1)R]^2 \phi(R)$$
$$+ \frac{k(1 - R)}{3}\left\{(1 + \beta + \beta^2) + (\beta - 1)(\beta + 2)R + (\beta - 1)^2 R^2\right\}. \qquad (7.46)$$

7.4   Integral formulation by Green's function

In this section we follow Section 3.3 and deduce an integral formulation of the problem by means of the Green's function associated with the operator appearing in the right-hand side of Eq. (7.1). In the usual way we introduce a function $w(r, t)$ which is the difference between the actual solution and the pseudo-steady-state solution, thus

$$w(r, t) = T(r, t) - T_{pss}(r, t), \qquad (7.47)$$

where $T_{pss}(r, t)$ is given by (1.77) and $R(t)$ in that equation denotes the actual reaction front rather than the pseudo-steady-state boundary. From (7.1), (7.2) and (7.47) we obtain the following self-adjoint problem with homogeneous boundary conditions, namely

$$\frac{\partial}{\partial r}\left(r^2 \frac{\partial w}{\partial r}\right) - k^2 r^2 w = r^2 \frac{\partial T}{\partial t}, \qquad R(t) < r < 1, \qquad (7.48)$$

$$w(R(t), t) = 0, \qquad w(1, t) + \beta \frac{\partial w}{\partial r}(1, t) = 0. \qquad (7.49)$$

As described in Section 3.3 we may deduce the following symmetric Green's function

$$G(r, \xi, t) = -\frac{\phi(r) \sinh k(\xi - R)}{kr\xi\phi(R)}, \qquad R \le \xi \le r,$$

$$= -\frac{\phi(\xi) \sinh k(r - R)}{kr\xi\phi(R)}, \qquad r \le \xi \le 1, \qquad (7.50)$$

where $\phi(r)$ is defined by (7.25) and we have

$$T(r, t) = T_{pss}(r, t) + \int_R^1 G(r, \xi, t)\xi^2 \frac{\partial T}{\partial t}(\xi, t)\,d\xi. \qquad (7.51)$$

From (7.50), (7.51) and the Stefan condition (7.3) we may deduce

$$1 = \frac{d}{dt}\left(\frac{1}{k}\int_R^1 \xi\phi(\xi)[\alpha + T(\xi, t)]\,d\xi\right),$$ (7.52)

which evidently may be integrated to yield the following integral formula for the boundary motion:

$$t = \frac{1}{k}\int_R^1 \xi\phi(\xi)[\alpha + T(\xi, t)]\,d\xi.$$ (7.53)

On multiplying T_{pss} in (7.51) by the right-hand side of (7.52) (that is, unity) and simplifying the result we may deduce

$$T(r, t) = \frac{1}{r}\frac{\partial}{\partial t}\left(\int_{R(t)}^r \xi\frac{\sinh k(r - \xi)}{k}[\alpha + T(\xi, t)]\,d\xi\right),$$ (7.54)

and (7.53) and (7.54) constitute the basic integral formulae for the problem (7.1)–(7.3). We observe that (3.23) and (3.21) are recovered from (7.53) and (7.54), respectively, in the limit k tending to zero.

In terms of $v(r, t)$ defined by (7.6) and using (r, R) as independent variables we may readily deduce, by dividing (7.54) by (7.52), the following integral formulation

$$v^\dagger(r, R) = \frac{\dfrac{\partial}{\partial R}\left(\displaystyle\int_R^r \sinh k(r - \xi)[\alpha\xi + v^\dagger(\xi, R)]\,d\xi\right)}{\dfrac{\partial}{\partial R}\left(\displaystyle\int_R^1 \phi(\xi)[\alpha\xi + v^\dagger(\xi, R)]\,d\xi\right)}.$$ (7.55)

Thus, from this equation and (7.53) we may formulate the iterative scheme

$$v_{n+1}(r, R) = \frac{\dfrac{\partial}{\partial R}\left(\displaystyle\int_R^r \sinh k(r - \xi)[\alpha\xi + v_n(\xi, R)]\,d\xi\right)}{\dfrac{\partial}{\partial R}\left(\displaystyle\int_R^1 \phi(\xi)[\alpha\xi + v_n(\xi, R)]\,d\xi\right)},$$ (7.56)

$$t_{n+1}(R) = \frac{1}{k}\int_R^1 \phi(\xi)[\alpha\xi + v_n(\xi, R)]\,d\xi.$$ (7.57)

With v_{pss} as the initial estimate and on noting the equation

$$\frac{\partial}{\partial R}v_{pss}^\dagger(r, R) = -\frac{k\phi(r)}{\phi(R)^2},$$ (7.58)

we find that $v_1(r, R)$ becomes

$$v_1(r, R) = \frac{\alpha R \sinh k(r - R) + [k/\phi(R)^2] \int_R^r \phi(\xi) \sinh k(r - \xi)\, d\xi}{\alpha R \phi(R) + [k/\phi(R)^2] \int_R^1 \phi(\xi)^2\, d\xi}.$$

(7.59)

The integral in the numerator of (7.59) can be evaluated with the substitution $\eta = k(r - \xi)$ and then expanding the sinh and cosh in $\phi(\xi)$ in terms of $\sinh \eta$ and $\cosh \eta$. On evaluating the resulting elementary integrals and simplifying the result we obtain

$$\int_R^r \phi(\xi) \sinh k(r - \xi)\, d\xi = \frac{1}{2k^2}\left\{\phi'(r)k(r - R) - \phi'(R) \sinh k(r - R)\right\}.$$

(7.60)

The integral in the denominator of (7.59) is most easily evaluated by noting that $\phi(r)$ defined by (7.25) satisfies the differential equation

$$\phi''(r) = k^2 \phi(r),$$

(7.61)

and that

$$\phi'(r)^2 = k^2 \phi(r)^2 + \phi'(1)^2 - k^2 \phi(1)^2.$$

(7.62)

Thus, using integration by parts we have

$$\int_R^1 \phi(\xi)^2\, d\xi = \frac{1}{k^2} \int_R^1 \phi(\xi)\phi''(\xi)\, d\xi$$

$$= \frac{1}{k^2}\left\{\phi(1)\phi'(1) - \phi(R)\phi'(R)\right\} - \frac{1}{k^2} \int_R^1 \phi'(\xi)^2\, d\xi, \quad (7.63)$$

and from this equation and (7.62) we may deduce

$$\int_R^1 \phi(\xi)^2\, d\xi = \frac{1}{2k^2}\left\{\phi(1)\phi'(1) - \phi(R)\phi'(R)\right\}$$

$$- \frac{1}{2k^2}\left\{\phi'(1)^2 - k^2\phi(1)^2\right\}(1 - R).$$

(7.64)

Using these integrals we find from (7.59) that $v_1(r, R)$ becomes

$$v_1(r, R) = v_{\text{pss}}^\dagger(r, R) + \frac{1}{\Delta_2(R)}\left\{\phi(R)\phi'(r)k(r - R)\right.$$

$$\left. - k^2\{k^2\beta^2(1 - R) + (\beta - 1)[1 + (\beta - 1)R]\} \sinh k(r - R)\right\},$$

(7.65)

where $\Delta_2(R)$ is defined by

$$\Delta_2(R) = 2\alpha R k \phi(R)^3$$
$$- \{\phi(R)\phi'(R) - k^2\{k^2\beta^2(1-R) + (\beta-1)[1+(\beta-1)R]\}\}.$$
(7.66)

With v_{pss} as the initial estimate for (7.57) and using the integral

$$\int_R^1 \phi(\xi) \sinh k(\xi - R)\, d\xi$$
$$= -\frac{1}{2k}\left\{(1-R)\phi'(R) + (1-\beta)\sinh k(1-R)\right\}, \quad (7.67)$$

which is evaluated in a similar manner to (7.60) but with the substitution $\eta = k(\xi - R)$, we find that $t_1(R)$ is simply the order one corrected motion given by the first two terms of (7.27), that is

$$t_1(R) = t_{\mathrm{pss}}(R)$$
$$+\left\{\frac{\begin{array}{c}k(1-R)[(1-\beta)\cosh k(1-R) + \beta k \sinh k(1-R)]\\ -(1-\beta)\sinh k(1-R)\end{array}}{2k^2[(1-\beta)\sinh k(1-R) + \beta k \cosh k(1-R)]}\right\}.$$
(7.68)

Moreover, we have the following estimate for the time to complete solidification

$$t_1(0) = t_{\mathrm{pssc}} + \left\{\frac{k[(1-\beta)\cosh k + \beta k \sinh k] - (1-\beta)\sinh k}{2k^2[(1-\beta)\sinh k + \beta k \cosh k]}\right\},$$
(7.69)

where t_{pssc} is given by (1.79).

The calculations for $v_2(r, R)$ are considerably more complicated and consequently we do not pursue the matter further. However, on noting the integral

$$\int_R^1 \phi(\xi)\phi'(\xi)(\xi - R)\, d\xi = \frac{1}{2}\left\{(1-R)\phi(1)^2 - \int_R^1 \phi(\xi)^2\, d\xi\right\}, \quad (7.70)$$

we may deduce from (7.57), (7.65) and (7.67) the following expression for $t_2(R)$, thus

$$t_2(R) = t_1(R) + \frac{k\phi(R)}{2\Delta_2(R)}\left\{(1-R)\phi(1)^2 - \int_R^1 \phi(\xi)^2\, d\xi\right\}$$
$$+\frac{k}{2\Delta_2(R)}\left\{\begin{array}{c}\{k^2\beta^2(1-R) + (\beta-1)[1+(\beta-1)R]\}\\ \times\{(1-R)\phi'(R) + (1-\beta)\sinh k(1-R)\}\end{array}\right\}, \quad (7.71)$$

which on using the integral (7.64) appears not to simplify further.

7.5 Bounds for the motion of the boundary

In this section we make use of the inequalities

$$0 \le T(r, t) \le T_{\text{pss}}(r, t), \tag{7.72}$$

and

$$-\frac{dR}{dt} \le \frac{k}{\alpha R \phi(R)}, \tag{7.73}$$

to obtain upper and lower bounds on the boundary motion. The left-hand side of (7.72) is physically apparent while the right-hand side follows immediately from (7.50) and (7.51) and the physically meaningful assumptions

$$\beta \ge 0, \qquad \frac{\partial T}{\partial t} \ge 0. \tag{7.74}$$

The upper bound on the velocity (7.73) arises from (7.52), namely

$$1 = -\frac{\alpha}{k} R\phi(R) \frac{dR}{dt} + \frac{1}{k} \int_R^1 \xi \phi(\xi) \frac{\partial T}{\partial t} (\xi, t) \, d\xi, \tag{7.75}$$

and the inequality $(7.74)_2$. From (7.53) and (7.72) we have immediately

$$t_{\text{pss}}(R) \le t(R) \le t_1(R), \tag{7.76}$$

where $t_{\text{pss}}(R)$ is given by (1.78) and $t_1(R)$ is the order one corrected boundary motion given by (7.68). We may also show that $\bar{t}_1(R)$ given by (7.41) constitutes an upper bound to the boundary motion. From (7.34) and the right-hand side of (7.72) we have

$$I(R) \le I_{\text{pss}}(R), \tag{7.77}$$

and therefore

$$\frac{1}{1 - k^2 I(R)} \le \frac{1}{1 - k^2 I_{\text{pss}}(R)}. \tag{7.78}$$

From this inequality, $(7.74)_1$ and (7.36) we have

$$t(R) \le \bar{t}_1(R). \tag{7.79}$$

We observe that for β zero numerical results given in Hill (1984a) indicate that $t_1(R)$ is marginally superior to $\bar{t}_1(R)$.

As described in Section 3.6 we may utilize (7.73) to improve the pseudo-steady-state lower bound. From (7.53) and (7.54) we have

$$t = t_{\text{pss}}(R) + \frac{1}{k^2} \frac{d}{dt} \int_R^1 \int_R^\xi \eta \phi(\xi) \sinh k(\xi - \eta)[\alpha + T(\eta, t)] \, d\eta \, d\xi,$$

$$\tag{7.80}$$

and in the usual way on integrating with respect to time, using (7.73) and $T \geq 0$ we may deduce

$$t^2 \geq t_{pss}^2(R) + \frac{2\alpha}{k^2} \int_R^1 \int_R^\xi \eta \phi(\xi) \sinh k(\xi - \eta) \, d\eta \, d\xi. \qquad (7.81)$$

On evaluating the double integral we obtain

$$t^2 \geq t_{pss}^2(R) + \frac{\alpha}{k^5} \Big\{ (2R - 1)\phi'(R) + [k^2 R(1 - R) - 3]\phi(R)$$

$$+ k[\beta + (\beta - 1)R] \cosh k(1 - R) + 2k \Big\}, \qquad (7.82)$$

which simplifies to give

$$t^2 \geq t_{pss}^2(R)$$

$$+ \frac{\alpha}{k^5} \Big\{ k[\beta k^2 R(1 - R) + (1 - 3R) + 3\beta(R - 1)] \cosh k(1 - R) + 2k$$

$$+ [k^2 R(1 - R) + \beta k^2 (R^2 - 3R + 1) - 3(1 - \beta)] \sinh k(1 - R) \Big\}.$$

$$(7.83)$$

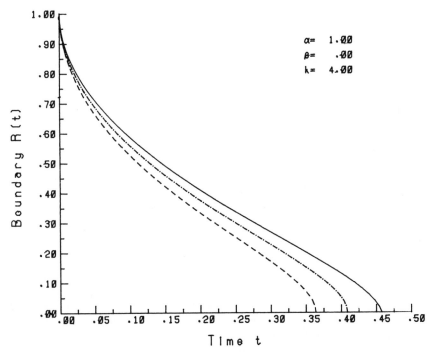

Figure 7.1 Variation of boundary motions for various bounds for $\alpha = 1.0$, β zero and $k = 4.0$ (– – – – pseudo-steady-state lower bound (1.78), ·–·–·– improved lower bound (7.83), ——— upper bound (7.68))

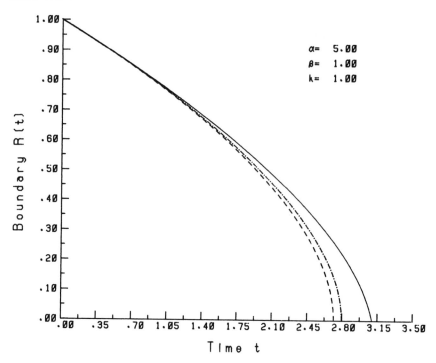

Figure 7.2 Variation of boundary motions for various bounds for $\alpha = 5.0$, $\beta = 1.0$ and $k = 1.0$ (– – – – pseudo-steady-state lower bound (1.78), ·—·—·— improved lower bound (7.83), ——— upper bound (7.68))

Thus, in particular, we have for the time to complete reaction

$$t_c^2 \geq t_{pssc}^2 + \frac{\alpha}{k^5}\left\{(\beta k^2 + 3\beta - 3)\sinh k + (1 - 3\beta)k\cosh k + 2k\right\}. \quad (7.84)$$

Figures 7.1 and 7.2 show the pseudo-steady-state motion $t_{pss}(R)$, the improved lower bound (7.83) and the order one corrected motion $t_1(R)$ for two sets of values of α, β and k. We observe that increasing any of α, β or k results in both a decrease of the velocity of the reaction front and a tightening of the bounds. This is expected on physical grounds, since increasing α corresponds to increasing the quantity of reacting solute required to advance the moving front a given distance. Increasing k corresponds to an increase in the rate at which the solute is consumed by the slow reaction and therefore decreases the solute flux at the moving interface. Further increasing β represents an increase in the mass transfer coefficient and consequently a reduction in the surface flux of solute into the particle. Clearly, the bounds are quite adequate for practical purposes and some average of them would represent an accurate approximation to the boundary motion, especially for large values of α or k. The problem of this section serves as a good example of the usefulness of such simple bounding techniques, since to obtain equivalent accuracy from either a semi-analytic series technique or a numerical scheme would involve considerably more analysis or computing time. In the following section, for completeness, we close this chapter with a derivation of the formal series solution for the problem (7.1)–(7.3), from which we may deduce the unsolved integral equation (7.103) for the inverse boundary motion $R^{-1}(r)$.

7.6 Formal series solution and related integral equation

We first show that a formal solution of (7.1), (7.2)$_2$ and (7.3)$_1$ is given by

$$T(r, t) = \frac{\alpha}{r} \frac{\partial}{\partial t} \sum_{n=1}^{\infty} \left(\delta + \frac{\partial}{\partial t} \right)^{n-1} \left\{ \frac{(r - R)^{2n}}{(2n + 1)!} (r + 2nR) \right\}, \tag{7.85}$$

where $\delta \equiv k^2$. This equation is the appropriate generalization of (4.18) and may be verified as a formal solution of (7.1) by writing (7.1) as

$$\left(\delta + \frac{\partial}{\partial t} \right)(rT) = \frac{\partial^2}{\partial r^2}(rT), \tag{7.86}$$

and verifying that both sides of the equation yield the same result. Further, (7.2)$_2$ and (7.3)$_1$ may be confirmed directly. The following derivation of (7.85) from the integral formula (7.54) may be of interest.

We first establish the following basic identity for $n \geq 0$:

$$\int_{\xi}^{r} \frac{\sinh k(r - \eta)}{k} \frac{\partial^n}{\partial \delta^n} \left[\frac{\sinh k(\eta - \xi)}{k} \right] d\eta$$

$$= \frac{1}{(n + 1)} \frac{\partial^{n+1}}{\partial \delta^{n+1}} \left[\frac{\sinh k(r - \xi)}{k} \right], \tag{7.87}$$

which is most easily proved by induction. The case n zero follows immediately using the elementary integral,

$$\int_{\xi}^{r} \sinh k(r - \eta) \sinh k(\eta - \xi) \, d\eta$$

$$= \frac{1}{2} \left\{ (r - \xi) \cosh k(r - \xi) - \frac{\sinh k(r - \xi)}{k} \right\}. \tag{7.88}$$

Assuming true for n and with I_n defined by

$$I_n = \int_{\xi}^{r} \frac{\sinh k(r - \eta)}{k} \frac{\partial^n}{\partial \delta^n} \left[\frac{\sinh k(\eta - \xi)}{k} \right] d\eta, \tag{7.89}$$

we have

$$I_{n+1} = \int_{\xi}^{r} \frac{\sinh k(r - \eta)}{k} \frac{\partial^{n+1}}{\partial \delta^{n+1}} \left[\frac{\sinh k(\eta - \xi)}{k} \right] d\eta$$

$$= (n + 1) \int_{\xi}^{r} \int_{\xi}^{\eta} \frac{\sinh k(r - \eta)}{k} \frac{\sinh k(\eta - \sigma)}{k}$$

$$\times \frac{\partial^{n}}{\partial \delta^{n}} \left[\frac{\sinh k(\sigma - \xi)}{k} \right] d\sigma \, d\xi$$

$$= (n + 1) \int_{\xi}^{r} \frac{\partial}{\partial \delta} \left[\frac{\sinh k(r - \sigma)}{k} \right] \frac{\partial^{n}}{\partial \delta^{n}} \left[\frac{\sinh k(\sigma - \xi)}{k} \right] d\sigma, \quad (7.90)$$

where the final line follows on changing orders of integration and utilizing the identity for $n = 1$. Thus, from (7.90) we have

$$I_{n+1} = (n + 1) \left\{ \frac{\partial I_n}{\partial \delta} - I_{n+1} \right\}, \tag{7.91}$$

from which we obtain

$$I_{n+1} = \left(\frac{n + 1}{n + 2} \right) \frac{\partial I_n}{\partial \delta}, \tag{7.92}$$

and the identity is established for $n + 1$ and the proof by induction is complete

Now as described in Chapter 4 we may on repeated application of (7.54), making use of (7.87), establish the following:

$$T(r, t) = \frac{\alpha}{r} \frac{\partial}{\partial t} \sum_{n=1}^{N} \frac{1}{(n - 1)!} \frac{\partial^{n-1}}{\partial t^{n-1}} \left(\int_{R}^{r} \xi \frac{\partial^{n-1}}{\partial \delta^{n-1}} \left[\frac{\sinh k(r - \xi)}{k} \right] d\xi \right)$$

$$+ R_N, \tag{7.93}$$

where the remainder R_N is given by

$$R_N = \frac{1}{r(N - 1)!} \frac{\partial^N}{\partial t^N} \left(\int_{R}^{r} \xi \frac{\partial^{N-1}}{\partial \delta^{N-1}} \left[\frac{\sinh k(r - \xi)}{k} \right] T(\xi, t) \, d\xi \right). \tag{7.94}$$

Thus, assuming R_N tends to zero as N tends to infinity we obtain

$$T(r, t) = \frac{\alpha}{r} \frac{\partial}{\partial t} \sum_{n=0}^{\infty} \frac{1}{n!} \frac{\partial^n}{\partial t^n} \left(\int_{R}^{r} \xi \frac{\partial^n}{\partial \delta^n} \left[\frac{\sinh k(r - \xi)}{k} \right] d\xi \right), \tag{7.95}$$

and on interchanging orders of integration and partial differentiation with respect to δ and using

$$\int_{R}^{r} \xi \frac{\sinh k(r - \xi)}{k} d\xi = \sum_{m=0}^{\infty} \delta^m \frac{(r - R)^{2m+2}}{(2m + 2)!} \left\{ R + \frac{r - R}{2m + 3} \right\}, \tag{7.96}$$

we find that (7.95) becomes

$$T(r, t) = \frac{\alpha}{r} \frac{\partial}{\partial t} \sum_{n=0}^{\infty} \frac{\partial^n}{\partial t^n} \sum_{m=n}^{\infty} \binom{m}{n} \delta^{m-n} \frac{(r-R)^{2m+2}}{(2m+2)!} \left\{ R + \frac{r-R}{2m+3} \right\}.$$

(7.97)

On changing orders of summation we find

$$T(r, t) = \frac{\alpha}{r} \frac{\partial}{\partial t} \sum_{m=0}^{\infty} \delta^m \left(1 + \frac{1}{\delta} \frac{\partial}{\partial t} \right)^m \frac{(r-R)^{2m+2}}{(2m+2)!} \left\{ R + \frac{r-R}{2m+3} \right\},$$

(7.98)

and (7.85) follows immediately from this equation on setting $n = m + 1$.

Equation (7.85) constitutes a formal solution to the problem (7.1)–(7.3) where the boundary motion $R(t)$ is determined from the surface condition (7.2)$_1$. Now, on using (7.85) in (7.2)$_1$ and performing one time integration we readily obtain

$$\frac{t}{\alpha} = \sum_{n=1}^{\infty} \left(\delta + \frac{d}{dt} \right)^{n-1} \left\{ \frac{(1-R)^{2n}}{(2n+1)!} (1 + 2nR) \right.$$

$$\left. + 2n\beta \frac{(1-R)^{2n-1}}{(2n+1)!} [R^2 + (2n-1)R + 1] \right\}.$$

(7.99)

Assuming that $f(t)$ is any function such that (4.31) holds we may show by induction

$$\int_0^{\infty} e^{-pt} \left(\delta + \frac{d}{dt} \right)^n f(t) \, dt = (\delta + p)^n \int_0^{\infty} e^{-pt} f(t) \, dt,$$

(7.100)

and thus on taking the Laplace transform of (7.99) we have

$$\int_0^{\infty} e^{-pt} \left\{ [(1 - \beta) + \beta sR] \frac{\sinh s^{1/2}(1-R)}{s^{1/2}} \right.$$

$$\left. + [(1 - \beta)R + \beta] \cosh s^{1/2}(1-R) - 1 \right\} dt = \frac{s}{\alpha p^2}, \quad (7.101)$$

where s denotes $\delta + p$. By integration by parts we obtain

$$\int_0^{\infty} e^{-pt} R \left\{ (1 - \beta) \frac{\sinh s^{1/2}(1-R)}{s^{1/2}} + \beta \cosh s^{1/2}(1-R) \right\} \frac{dR}{dt} \, dt$$

$$= -\frac{1}{\alpha p}, \quad (7.102)$$

and consequently with the change of variable $\xi = R(t)$ we find that

$R^{-1}(r)$ is determined as a solution of the integral equation

$$\int_0^1 e^{-pR^{-1}(\xi)}\xi\left\{(1-\beta)\frac{\sinh s^{1/2}(1-\xi)}{s^{1/2}}+\beta\cosh s^{1/2}(1-\xi)\right\}d\xi$$

$$=\frac{1}{\alpha p}, \quad (7.103)$$

where $s \equiv k^2 + p$. Evidently (7.103) can also be derived via the modified Laplace transforms as described in Chapter 4. These details as well as results for slab and cylindrical geometries can be found in Hill (1984b) and Dewynne and Hill (1985).

7.7 Additional symbols used

a	radius of spherical particle, (7.4)
$A(\tau)$, $B(\tau)$	arbitrary functions, (7.21)
D	diffusivity of the chemical reactant through the fluid, (7.4)
$G(r, \xi, t)$	Green's function, (7.50)
h_D	surface mass transfer coefficient, (7.5)
$I(R)$	integral defined by (7.34)
$I_{pss}(R)$	pseudo-steady-state $I(R)$, (7.39)
$I_1(R)$	integral $I(R)$ arising from $\bar{v}_1(r, R)$, (7.44)
$J(r, R)$	integral defined by (7.37)
$J_{pss}(r, R)$	pseudo-steady-state $J(r, R)$, (7.40)
k	constant $(a^2 k_1/D)^{1/2}$, (7.5)
k_1	first-order rate constant, (7.5)
p	Laplace transform variable, (7.100)
s	denotes $k^2 + p$, (7.100)
T_0	constant surface concentration of fluid reactant, (7.4)
$t_n(R)$	iterations for boundary motion using Green's function formulation, (7.57)
$t_1(R)$	first iteration from v_{pss}, (7.68)
$t_2(R)$	second iteration from v_{pss}, (7.71)
$\bar{t}_n(R)$	iterations for boundary motion using direct integration formulation, (7.36)
$\bar{t}_1(R)$	first iteration from v_{pss}, (7.41)
$\bar{t}_2(R)$	second iteration from v_{pss}, (7.44)
$u(\rho, \tau)$	rT in terms of boundary-fixing variables, (7.12)
$u_n(\rho, \tau)$	coefficients in large α expansion, (7.18)
$v(r, t)$	rT with r and t as independent variables, (7.6)
$\bar{v}_1(r, t)$	denotes arbitrary solution of one-dimensional heat equation, (7.10)

$v_n(r, R)$ iterations for rT using Green's function formulation, (7.56)
$v_1(r, R)$ first iteration from v_{pss}, (7.65)
$\bar{v}_n(r, R)$ iterations for rT using direct integration formulation, (7.38)
$\bar{v}_1(r, R)$ first iteration from v_{pss}, (7.42)
$w(r, t)$ difference $T - T_{pss}$, (7.47)

Greek symbols

δ denotes k^2, (7.85)
$\Delta_1(R)$ function defined by (7.46)
$\Delta_2(R)$ function defined by (7.66)
η integration variable
ρ boundary-fixing variable, (7.12)
ρ_1 density of solid, (7.5)
τ boundary-fixing variable, (7.12)
$\tau_n(R)$ coefficients in large α expansion of motion, (7.27)
$\phi(r)$ function defined by (7.25)
ω stoichiometric coefficient for rapid reaction, (7.5)

Convention

Primes used denote differentiation with respect to the argument indicated.

Appendix 1

Langford's cylinder functions $c_n(z, z_0)$ and $e_n(z, z_0)$

In this appendix we verify the basic integral formulae (4.2)–(4.4) and deduce general expressions for Langford's cylinder functions $c_n(z, z_0)$ and $e_n(z, z_0)$. From the appendix of Langford (1967b) it is apparent that for $n \geq 1$ $c_n(z, z_0)$ and $e_n(z, z_0)$ are generated recursively from

$$\frac{\partial}{\partial z}\left(z \frac{\partial f_n}{\partial z}(z, z_0)\right) = f_{n-1}(z, z_0), \tag{A1.1}$$

$$f_n(z_0, z_0) = \frac{\partial f_n}{\partial z}(z_0, z_0) = 0, \tag{A1.2}$$

together with

$$c_0(z, z_0) = 1, \qquad e_0(z, z_0) = \log\left(\frac{z}{z_0}\right), \tag{A1.3}$$

respectively. On integrating (A1.1) twice, interchanging orders of integration and using (A1.2) we obtain

$$f_n(z, z_0) = \int_{z_0}^{z} \log\left(\frac{z}{\omega}\right) f_{n-1}(\omega, z_0) \, d\omega \qquad (n \geq 1), \tag{A1.4}$$

and thus (4.2) and (4.3) are established. We prove (4.4) by induction. By direct integration of $e_0(z, \omega)$ we may readily verify (4.4) for $n = 1$. Assuming true for n we consider

$$\int_{z_0}^{z} e_n(z, \omega) \, d\omega = \int_{z_0}^{z} \int_{\omega}^{z} \log\left(\frac{z}{\eta}\right) e_{n-1}(\eta, \omega) \, d\eta \, d\omega$$

$$= \int_{z_0}^{z} \int_{z_0}^{\eta} \log\left(\frac{z}{\eta}\right) e_{n-1}(\eta, \omega) \, d\omega \, d\eta$$

$$= \int_{z_0}^{z} \log\left(\frac{z}{\eta}\right) c_n(\eta, z_0) \, d\eta$$

$$= c_{n+1}(z, z_0), \tag{A1.5}$$

where the various steps follow from (4.3), (4.4) assumed true for n and finally (4.2). Thus we have established (4.4) by induction.

We now proceed to deduce explicit expressions for $c_n(z, z_0)$ and $e_n(z, z_0)$. By inspection of (2.24) and (2.25) and by induction we may show that in both cases the functions take the form

$$f_n(z, z_0) = z_0^n g_n\left(\frac{z}{z_0}\right) \qquad (n \geq 1), \tag{A1.6}$$

so that from (A1.1) and (A1.2) we have, for $n \geq 1$,

$$\xi g_n''(\xi) + g_n'(\xi) = g_{n-1}(\xi), \tag{A1.7}$$
$$g_n(1) = g_n'(1) = 0, \tag{A1.8}$$

where primes here denote differentiation with respect to $\xi = z/z_0$. Thus, if we define the generating function

$$G(\xi, s) = \sum_{n=0}^{\infty} g_n(\xi) s^n, \tag{A1.9}$$

then, since in both cases we have

$$\xi g_0''(\xi) + g_0'(\xi) = 0, \tag{A1.10}$$

we obtain from (A1.7) and (A1.9)

$$\xi \frac{\partial^2 G}{\partial \xi^2} + \frac{\partial G}{\partial \xi} - sG = 0. \tag{A1.11}$$

For the functions $c_n(z, z_0)$ the generating function $C(\xi, s)$ is such that

$$C(1, s) = 1, \qquad \frac{\partial C}{\partial \xi}(1, s) = 0, \tag{A1.12}$$

and it is not difficult to establish that the required solution of (A1.11) is

$$C(\xi, s) = 2\sqrt{s} \left\{ I_0[2\sqrt{(s\xi)}]K_1(2\sqrt{s}) + K_0[2\sqrt{(s\xi)}]I_1(2\sqrt{s}) \right\}, \tag{A1.13}$$

where I_0, I_1, K_0 and K_1 denote the usual modified Bessel functions. Similarly, if $E(\xi, s)$ denotes the generating function for the functions $e_n(z, z_0)$ then we have

$$E(1, s) = 0, \qquad \frac{\partial E}{\partial \xi}(1, s) = 1, \tag{A1.14}$$

and the appropriate solution of (A1.11) is

$$E(\xi, s) = 2\left\{ I_0[2\sqrt{(s\xi)}]K_0(2\sqrt{s}) - K_0[2\sqrt{(s\xi)}]I_0(2\sqrt{s}) \right\}. \tag{A1.15}$$

On utilizing the standard series for Bessel functions, namely

$$I_0(x) = \sum_{n=0}^{\infty} \frac{(x/2)^{2n}}{(n!)^2}, \qquad I_1(x) = \sum_{n=0}^{\infty} \frac{(x/2)^{2n+1}}{n!\,(n+1)!},$$

$$K_0(x) = -\left(\gamma + \log\left(\frac{x}{2}\right)\right)I_0(x) + \sum_{n=0}^{\infty}\left\{1 + \frac{1}{2} + \ldots + \frac{1}{n}\right\}\frac{(x/2)^{2n}}{(n!)^2},$$

$$K_1(x) = \left(\gamma + \log\left(\frac{x}{2}\right)\right)I_1(x) + \frac{1}{x}$$

$$- \sum_{n=0}^{\infty}\left[\left\{1 + \frac{1}{2} + \ldots + \frac{1}{n}\right\} + \frac{1}{2(n+1)}\right]\frac{(x/2)^{2n+1}}{n!\,(n+1)!}, \qquad \text{(A1.16)}$$

where here $\gamma = 0.5772157\ldots$ denotes Euler's constant and $\{1 + \frac{1}{2} + \ldots + 1/n\}$ is taken to be zero when n is zero, we may expand (A1.13) and (A1.15) to deduce the following expressions for $c_n(z, z_0)$ and $e_n(z, z_0)$ for $n \geq 1$:

$$\frac{c_n(z, z_0)}{z_0^n} = \frac{2}{[(n-1)!]^2}\sum_{j=0}^{n-1}\binom{n-1}{j}^2\left\{1 + \frac{1}{2} + \ldots + \frac{1}{n-j-1}\right\}$$

$$\times \left\{\frac{\xi^{n-j-1}}{j+1} - \frac{\xi^j}{n-j}\right\} + \frac{\xi^n}{(n!)^2}$$

$$- \frac{1}{(n!)^2}\sum_{j=0}^{n-1}\binom{n}{j}^2 \xi^j[1 + (n-j)\log\xi], \qquad \text{(A1.17)}$$

$$\frac{e_n(z, z_0)}{z_0^n} = \frac{2}{(n!)^2}\sum_{j=0}^{n}\binom{n}{j}^2\left\{1 + \frac{1}{2} + \cdots + \frac{1}{j}\right\}[\xi^{n-j} - \xi^j]$$

$$+ \frac{1}{(n!)^2}\sum_{j=0}^{n}\binom{n}{j}^2 \xi^{n-j}\log\xi, \qquad \text{(A1.18)}$$

where $\xi = z/z_0$. We may confirm that the first four functions given by (2.24) and (2.25) arise from the above general formulae. The results given in this appendix are due to Hill and Dewynne (1985).

A1.1 Additional symbols used

$C(\xi, s)$ generating function of $c_n(z, z_0)/z_0^n$, (A1.13)
$E(\xi, s)$ generating function of $e_n(z, z_0)/z_0^n$, (A1.15)
$f_n(z, z_0)$ denotes either $c_n(z, z_0)$ or $e_n(z, z_0)$, (A1.1)
$g_n(\xi)$ defined by (A1.6)

$G(\xi, s)$ denotes either $C(\xi, s)$ or $E(\xi, s)$, (A1.9)

z, z_0 denote $r^2/4$, $r_0^2/4$, respectively

Greek symbols

η, ω integration variables

ξ defined as z/z_0

Appendix 2

Evaluating $A_1(\rho)$, $A_2(\rho)$ and $A_3'(1)$ for spherical solidification

In this appendix we give details of the evaluation of $A_1(\rho)$, $A_2(\rho)$ and $A_3'(1)$ arising in the series solution (6.4) for the problem of spherical solidification (2.4)–(2.6), with prescribed surface temperature (that is, β zero) and with variables defined by (6.2).

A2.1 Solution for $A_1(\rho)$

Since $A_0(\rho)$ is given by (6.39) (also Eq. (1.49)) we have from (6.52) with $n = 1$

$$f_1(\rho) = \alpha\gamma\left(\gamma - \frac{A_1'(1)}{\alpha}\right)\rho \, e^{\gamma(1-\rho^2)/2}, \tag{A2.1}$$

so that from (6.43) and (6.65)$_1$ we find that (6.64) becomes

$$A_1'(1) = \alpha\gamma\left(\gamma - \frac{A_1'(1)}{\alpha}\right)\int_0^1 \xi^2 \, d\xi. \tag{A2.2}$$

Thus it follows that $A_1'(1)$ is given by (6.68)$_2$ and $f_1(\rho)$ becomes

$$f_1(\rho) = \frac{3\alpha\gamma^2\rho}{(\gamma + 3)} e^{\gamma(1-\rho^2)/2}. \tag{A2.3}$$

In order to evaluate $A_1(\rho)$ we use (6.62) with C_{11} zero and C_{12} given by (6.63). From (6.43) and (A2.3) we have

$$\frac{f_1(\xi)}{\omega(\xi)} = -\frac{3\alpha\gamma^2}{(\gamma + 3)} e^{\gamma/2}\xi, \tag{A2.4}$$

while from (6.65)$_1$ and (6.65)$_2$ we have for $0 \le \xi \le \rho$

$$A_{11}(\rho)A_{12}(\xi) - A_{11}(\xi)A_{12}(\rho)$$
$$= \xi \, e^{-\gamma\rho^2/2} - \rho \, e^{-\gamma\xi^2/2} + \gamma\rho\xi\int_\xi^\rho e^{-\gamma\eta^2/2} \, d\eta. \tag{A2.5}$$

On performing the integrations we have from these two equations

$$\int_0^\rho \frac{f_1(\xi)}{\omega(\xi)} \left\{ A_{11}(\rho)A_{12}(\xi) - A_{11}(\xi)A_{12}(\rho) \right\} d\xi = -\frac{\alpha\gamma\rho}{(\gamma+3)} e^{\gamma(1-\rho^2)/2},$$

(A2.6)

from which it is apparent that the constant C_{12} is $\alpha\gamma/(\gamma+3)$ and (6.62) gives the expression $(6.66)_1$ for $A_1(\rho)$.

A2.2 Solution for $A_2(\rho)$

From (6.52) with $n = 2$ we have

$$f_2(\rho) = A_1'' + \frac{1}{\alpha}\left\{ A_1'(1)(\rho A_1' - A_1) + A_2'(1)\rho A_0' \right\},$$

(A2.7)

which on using expressions (6.39) and $(6.66)_1$ for $A_0(\rho)$ and $A_1(\rho)$, respectively, simplifies to give

$$f_2(\rho) = \left\{ \left[\frac{3\alpha\gamma}{\gamma+3} - A_2'(1) \right] - \frac{3\alpha\gamma^2\rho^2}{(\gamma+3)^2} \right\} \gamma\rho \, e^{\gamma(1-\rho^2)/2}.$$

(A2.8)

From this equation, (6.43) and $(6.65)_4$ we find that the basic equation (6.64) for the determination of $A_2'(1)$ becomes

$$A_2'(1) = \int_0^1 (A + B\xi^2) \left\{ \xi \, e^{-\gamma\xi^2/2} + (1 + \gamma\xi^2) \int_0^\xi e^{-\gamma\eta^2/2} \, d\eta \right\} \xi \, d\xi, \quad \text{(A2.9)}$$

where the constants A and B are defined by

$$A = \frac{\alpha\gamma^2 e^{\gamma/2}}{(\gamma+1+\alpha\gamma)} \left\{ \frac{3\alpha\gamma}{(\gamma+3)} - A_2'(1) \right\},$$

$$B = \frac{-3\alpha^2\gamma^4 e^{\gamma/2}}{(\gamma+1+\alpha\gamma)(\gamma+3)^2}.$$

(A2.10)

If we define $I(\rho)$ by

$$I(\rho) = \int_0^\rho e^{-\gamma\eta^2/2} \, d\eta,$$

(A2.11)

then (A2.9) becomes

$$A_2'(1) = \int_0^1 (A + B\xi^2) \left\{ \frac{d}{d\xi}[\xi I(\xi)] + \gamma\xi^2 I(\xi) \right\} \xi \, d\xi,$$

(A2.12)

which on integration by parts gives

$$A_2'(1) = \frac{e^{-\gamma/2}}{\alpha\gamma} \left\{ \frac{A}{4}(\gamma + 2) + \frac{B}{12}(2\gamma + 3) \right\} + \frac{A}{2}\bar{I}_1 + \frac{3B - \gamma A}{4}\bar{I}_2 - \frac{\gamma B}{6}\bar{I}_3,$$

(A2.13)

where the integrals $\bar{I}_1$, $\bar{I}_2$ and $\bar{I}_3$ are defined by

$$\bar{I}_1 = \int_0^1 \xi^2 e^{-\gamma\xi^2/2} \, d\xi = \frac{e^{-\gamma/2}}{\alpha\gamma^2}(1 - \alpha\gamma),$$

$$\bar{I}_2 = \int_0^1 \xi^4 e^{-\gamma\xi^2/2} \, d\xi = \frac{e^{-\gamma/2}}{\alpha\gamma^3}[3 - \alpha\gamma(\gamma + 3)],$$

(A2.14)

$$\bar{I}_3 = \int_0^1 \xi^6 e^{-\gamma\xi^2/2} \, d\xi = \frac{e^{-\gamma/2}}{\alpha\gamma^4}[15 - \alpha\gamma(\gamma^2 + 5\gamma + 15)].$$

From (A2.13) and (A2.14) we have

$$A_2'(1) = \frac{A \, e^{-\gamma/2}}{4\alpha\gamma^2}[(\gamma^2 + 2\gamma - 1) + \alpha\gamma(\gamma + 1)]$$

$$+ \frac{B \, e^{-\gamma/2}}{12\alpha\gamma^3}[(2\gamma^3 + 3\gamma^2 - 3) + \alpha\gamma(2\gamma^2 + \gamma + 3)],$$

(A2.15)

and on using (A2.10) this equation gives rise to the expression for $A_2'(1)$ given by (6.68)$_3$ and (6.69)$_1$.

In order to determine $A_2(\rho)$ explicitly we have from (6.43) and (A2.8)

$$\frac{f_2(\xi)}{\omega(\xi)} = -(a + b\xi^2)\xi \, e^{\gamma/2},$$

(A2.16)

where the constants a and b are defined by

$$a = \left[\frac{3\alpha\gamma}{(\gamma + 3)} - A_2'(1) \right]\gamma, \qquad b = -\frac{3\alpha\gamma^3}{(\gamma + 3)^2}.$$

(A2.17)

Further, from (6.65)$_3$ and (6.65)$_4$ we have for $0 \le \xi \le \rho$

$$A_{21}(\rho)A_{22}(\xi) - A_{21}(\xi)A_{22}(\rho)$$

$$= \frac{1}{2} \left\{ (1 + \gamma\rho^2)\xi \, e^{-\gamma\xi^2/2} - (1 + \gamma\xi^2)\rho \, e^{-\gamma\rho^2/2} \right.$$

$$\left. - (1 + \gamma\rho^2)(1 + \gamma\xi^2)\int_\xi^\rho e^{-\gamma\eta^2/2} \, d\eta \right\}.$$

(A2.18)

From (A2.16) and (A2.18) we may evaluate the integral in (6.62) for $n = 2$ with the aid of the following elementary integrals:

$$\int_0^\rho \xi^2 e^{-\gamma\xi^2/2} \, d\xi = \frac{1}{\gamma}\left\{I(\rho) - \rho\, e^{-\gamma\rho^2/2}\right\},$$

$$\int_0^\rho \xi^4 e^{-\gamma\xi^2/2} \, d\xi = \frac{1}{\gamma^2}\left\{3I(\rho) - \rho(3 + \gamma\rho^2)\, e^{-\gamma\rho^2/2}\right\}, \qquad (A2.19)$$

$$\int_0^\rho \xi^6 e^{-\gamma\xi^2/2} \, d\xi = \frac{1}{\gamma 3}\left\{15I(\rho) - \rho(15 + 5\gamma\rho^2 + \gamma^2\rho^4)\, e^{-\gamma\rho^2/2}\right\}.$$

After evaluating the integral in (6.62), namely

$$J(\rho) = \int_0^\rho \frac{f_2(\xi)}{\omega(\xi)}\left\{A_{21}(\rho)A_{22}(\xi) - A_{21}(\xi)A_{22}(\rho)\right\} d\xi, \qquad (A2.20)$$

then the solution for $A_2(\rho)$ becomes

$$A_2(\rho) = J(\rho) - \frac{J(1)}{A_{22}(1)} A_{22}(\rho), \qquad (A2.21)$$

and this is a useful equation in the calculation. After an extremely long and tedious calculation we may eventually simplify (A2.21) to give $(6.66)_2$.

A2.3 Solution for $A_3'(1)$

Due to the complexity of the analysis we only determine $A_3'(1)$ which is required for the determination of the boundary motion. From (6.52) with $n = 3$ we have

$$f_3(\rho) = A_2'' + \frac{1}{\alpha}\sum_{j=1}^3 A_j'(1)[\rho A_{3-j}' - (3-j)A_{3-j}], \qquad (A2.22)$$

which on using the expressions (6.39), $(6.66)_1$ and $(6.66)_2$ for $A_0(\rho)$, $A_1(\rho)$ and $A_2(\rho)$, respectively, simplifies to give

$$f_3(\rho) = (B_1 + B_2\rho^2 + B_3\rho^4)\rho\, e^{\gamma(1-\rho^2)/2} + B_4 I(\rho), \qquad (A2.23)$$

where $I(\rho)$ is given by (A2.11) and the constants B_1, B_2, B_3 and B_4 are defined by

$$B_1 = \frac{\alpha\gamma^2}{4(\gamma+3)^3}[3(4\gamma^2+21\gamma+30)-(4\gamma+9)\mu_1]-\gamma A_3'(1),$$

$$B_2 = \frac{-\alpha\gamma^3}{4(\gamma+3)^3}[3(7\gamma+20)-7\mu_1],$$

$$B_3 = \frac{3\alpha\gamma^4}{2(\gamma+3)^3},\qquad\qquad\qquad\qquad\qquad\text{(A2.24)}$$

$$B_4 = \frac{3\alpha^2\gamma^3\,e^{\gamma/2}(\gamma+6-\mu_1)}{2(\gamma+3)^3(\gamma+1+\alpha\gamma)},$$

where the constant μ_1 is defined by $(6.69)_1$. For $n=3$ the basic equation (6.64) for the determination of $A_3'(1)$ becomes on using $(6.65)_5$

$$A_3'(1) = \frac{1}{(\gamma+3)}\int_0^1 f_3(\xi)\,e^{-\gamma(1-\xi^2)/2}(3+\gamma\xi^2)\xi\,d\xi. \qquad\text{(A2.25)}$$

From (A2.23), (A2.25) and performing the integrations we may eventually deduce the expression for $A_3'(1)$ given by $(6.68)_4$ and (6.69).

A2.4 Additional symbols used

$A_1(\rho)$, $A_2(\rho)$, $A_3(\rho)$	coefficients in (6.4) for spherical solidification
$A_{11}(\rho)$, $A_{12}(\rho)$, $A_{21}(\rho)$, $A_{22}(\rho)$	linearly independent solutions of homogeneous equation in (6.15) for $n=1$ and $n=2$, (6.65)
a, b	constants defined by (A2.17)
A, B	constants defined by (A2.10)
B_1, B_2, B_3, B_4	constants defined by (A2.24)
$f_1(\rho)$, $f_2(\rho)$, $f_3(\rho)$	right-hand sides of (6.15) for $n=1$, 2 and 3, (6.52)
$\bar{I}_1$, $\bar{I}_2$, $\bar{I}_3$	integrals defined by (A2.14)
$I(\rho)$, $J(\rho)$	integrals defined by (A2.11) and (A2.20), respectively

Greek symbols

γ	positive root of transcendental equation (1.43)
η, ξ	integration variables

μ_1	constant defined by $(6.69)_1$
ρ	boundary-fixing transformation, $(6.2)_1$
$\omega(\rho)$	Wronskian defined by (6.43)

Convention

Primes denote differentiation with respect to ρ.

References

Books on moving-boundary problems

Albrecht, J., Collatz, L. and Hoffmann, K. H. (1982). *Numerical Treatment of Free-Boundary-Value Problems*. Birkhäuser Verlag, Boston.

Carslaw, H. S. and Jaeger, J. C. (1965). *Conduction of Heat in Solids*. 2nd Ed. Clarendon Press, Oxford.

Cohen, H. (1971). *Non-linear Diffusion Problems*. Studies in Mathematics, Vol. 7. Studies in Applied Mathematics (ed. A. H. Taub). Prentice-Hall, Englewood Cliffs, New Jersey.

Crank, J. (1964). *The Mathematics of Diffusion*. Clarendon Press, Oxford.

Crank, J. (1984). *Free- and Moving-Boundary Problems*. Clarendon Press, Oxford.

Elliott, C. M. and Ockendon, J. R. (1982). *Weak and Variational Methods for Moving-Boundary Problems*. Research Notes in Mathematics. Pitman, London.

Flemings, M. C. (1974). *Solidification Processing*. McGraw-Hill, New York.

Lewis, R. W. and Morgan, K. (1979). *Numerical Methods in Thermal Problems*. Pineridge Press, Swansea.

Ockendon, J. R. and Hodgkins, W. R. (1975). *Moving-Boundary Problems in Heat Flow and Diffusion*. Clarendon Press, Oxford.

Rubinstein, L. I. (1971). *The Stefan Problem*. Translations of Mathematical Monographs, Vol. 27. Am. Math. Soc., Providence, Rhode Island.

Wilson, D. G., Solomon, A. D. and Boggs, P. T. (1978). *Moving-Boundary Problems*. Academic Press, New York.

Review articles on moving-boundary problems

Bankoff, S. G. (1964). Heat conduction or diffusion with change of phase. *Adv. Chem. Engng.* **5,** 75–150.

Boley, B. A. (1972). Survey of recent developments in the fields of heat conduction in solids and thermo-elasticity. *Nuclear Engng. Design* **18,** 377–399.

Furzeland, R. M. (1977). A survey of the formulation and solution of free- and moving-boundary (Stefan) problems. Technical Report, Dept. of Mathematics, Brunel University, Uxbridge.

Goodman, T. R. (1964). Application of integral methods to transient non-linear heat transfer. *Adv. Heat Transfer* **1**, 51–122.

Mori, A. and Araki, K. (1976). Methods of analysis of the moving-boundary-surface problem. *Int. Chem. Engng.* **16**, 734–744.

Muehlbauer, J. C. and Sunderland, J. E. (1965). Heat conduction with freezing or melting. *Appl. Mech. Rev.* **18**, 951–959.

Rubinstein, L. (1979). The Stefan problem: comments on its present state. *J. Inst. Maths. Applics.* **24**, 259–277.

General references

Babu, D. K. and van Genuchten, M. Th. (1979). A similarity solution to a non-linear diffusion equation of the singular type. *Quart. Appl. Math.* **37**, 11–21.

Bell, G. E. (1978). A refinement of the heat balance integral method applied to a melting problem. *Int. J. Heat Mass Transfer* **21**, 1357–1362.

Bell, G. E. (1979). Solidification of a liquid about a cylindrical pipe. *Int. J. Heat Mass Transfer* **22**, 1681–1686.

Biot, M. A. (1957). New methods in heat-flow analysis with application to flight structures. *J. Aeronaut. Sci.* **24**, 857–873.

Biot, M. A. (1959). Further developments of new methods in heat-flow analysis. *J. Aerospace Sci.* **26**, 367–381.

Biot, M. A. and Daughaday, H. (1962). Variational analysis of ablation. *J. Aerospace Sci.* **29**, 227–229.

Bischoff, K. B. (1963). Accuracy of the peudo-steady-state approximation for moving-boundary diffusion problems. *Chem. Engng. Sci.* **18**, 711–713.

Bischoff, K. B. (1965). Further comments on the pseudo-steady-state approximation for moving-boundary diffusion problems. *Chem. Engng. Sci.* **20**, 783–784.

Bluman, G. W. (1974). Applications of the general similarity solution of the heat equation to boundary-value problems. *Quart. Appl. Math.* **31**, 403–415.

Boley, B. A. (1961). A method of heat conduction analysis of melting and solidifying slabs. *J. Math. Phys.* **40**, 300–313.

Boley, B. A. (1968). A general starting solution for melting and solidifying slabs. *Int. J. Engng. Sci.* **6**, 89–111.

Boley, B. A. and Yagoda, H. P. (1969). The starting solution for two-dimensional heat-conduction problems with change of phase. *Quart. Appl. Math.* **27**, 223–246.

Brillouin, M. (1931). Sur quelques problèmes non résolus de la physique mathematique classique propagation de la fusion. *Ann. Inst. H. Poincaré* **1**, 285–308.

Chalmers, B. (1954). Melting and freezing. *Trans. AIME, J. of Metals* **200**, 519–532.

Cho, S. H. (1975). An exact solution of the coupled phase-change problem in a porous medium. *Int. J. Heat Mass Transfer* **18**, 1139–1142.

Chuang, Y. K. and Ehrich, O. (1974). On the integral technique for spherical growth problems. *Int. J. Heat Mass Transfer* **17**, 945–953.

Chuang, Y. K. and Szekely, J. (1971). On the use of Green's functions for solving melting or solidification problems. *Int. J. Heat Mass Transfer* **14**, 1285–1294.

Chuang, Y. K. and Szekely, J. (1972). The use of Green's function for solving melting or solidification problems in the cylindrical coordinate system. *Int. J. Heat Mass Transfer* **15**, 1171–1174.

Chung, B. T. F. and Yeh, L. T. (1975). Solidification and melting of materials subject to convection and radiation. *J. Spacecraft and Rockets* **12**, 329–333.

Cross, M., Gibson, R. D. and Young, R. W. (1979). Pressure generation during the drying of a porous half-space. *Int. J. Heat Mass Transfer* **22**, 47–50.

Dana, P. R. and Wheelock, T. D. (1974). Kinetics of a moving-boundary ion-exchange process. *Ind. Engng. Chem. Fundam.* **13**, 20–26.

Danckwerts, P. V. (1950). Unsteady-state diffusion or heat conduction with moving boundary. *Trans. Faraday Soc.* **46**, 701–712.

Davis, G. B. and Hill, J. M. (1982). A moving-boundary problem for the sphere. *IMA J. Appl. Maths.* **29**, 99–111.

De Mey, G. (1977). A comment on 'An integral equation method for diffusion'. *Int. J. Heat Mass Transfer* **20**, 181–182.

Dewey, C. F., Schlesinger, S. I. and Sashkin, L. (1960). Temperature profiles in a finite solid with moving boundary. *J. Aerospace Sci.* **27**, 59–64.

Dewynne, J. N. and Hill, J. M. (1984). On an integral formulation for moving-boundary problems. *Quart. Appl. Math.* **41**, 443–456.

Dewynne, J. N. and Hill, M. J. (1985). Bounds for moving-boundary problems with two chemical reactions. *J. Nonlinear Anal.* **9**, 1293–1302.

Duck, P. W. and Riley, D. S. (1977). An extension of existing solidification results obtained from the heat balance integral method. *Int. J. Heat Mass Transfer* **20**, 297.

Duda, J. L., Malone, M. F. and Notter, R. H. (1975). Analysis of two-dimensional diffusion-controlled moving-boundary problems. *Int. J. Heat Mass Transfer* **18**, 901–910.

El-Genk, M. S. and Cronenberg, A. W. (1979). Some improvements to the solution of Stefan-like problems. *Int. J. Heat Mass Transfer* **22**, 167–170.

Elmer, S. L. (1932). Ice formation on pipe surfaces. *Refrigerating Engineering* **24**, 17–19.

Essoh, D. C. and Klinzing, G. E. (1980). Heat and mass diffusion with chemical reaction: A moving-boundary analysis in a particle. *A.I.Ch.E. J.* **26**, 869–871.

Field, A. L. (1927). Solidification of steel in the ingot mould. *Am. Soc. Metals* **11**, 264–276.

Gibson, R. E. (1958). A heat-conduction problem involving a specified moving boundary. *Quart. Appl. Math.* **16**, 426–430.

Glasser, D. and Kern, J. (1978). Bounds and approximate solutions to linear problems with non-linear boundary conditions: Solidification of a slab. *A.I.Ch.E. J.* **24**, 161–170.

Goodling, J. S. and Khader, M. S. (1975). Results of the numerical solution for outward solidification with flux boundary conditions. *ASME J. Heat Transfer* **94**, 307–309.

Goodman, T. R. (1958). The heat-balance integral and its application to problems involving a change of phase. *ASME Trans.* **80**, 335–342.

Goodman, T. R. (1961). The heat-balance integral—further considerations and refinements. *ASME J. Heat Transfer* **83**, 83–86.

Goodman, T. R. and Shea, J. J. (1960). The melting of finite slabs. *ASME J. Appl. Mech.* **27**, 16–24.

Grange, B. W., Viskanta, R. and Stevenson, W. H. (1976). Diffusion of heat and solute during freezing of salt solutions. *Int. J. Heat Mass Transfer* **19**, 373–384.

Grinberg, G. A. and Chekmareva, O. M. (1971). Motion of the phase interface in the Stefan problem. *Sov. Phys. Tech. Phys.* **15**, 1579–1583.

Grundy, R. E. (1979). Similarity solutions of the non-linear diffusion equation. *Quart. Appl. Math.* **37**, 259–280.

Gupta, L. N. (1974). An approximate solution of the generalized Stefan problem in a porous media. *Int. J. Heat Mass Transfer* **17**, 313–321.

Gutfinger, C. and Chen, W. H. (1969). Heat transfer with a moving boundary—application to fluidized bed coating. *Int. J. Heat Mass Transfer* **12**, 1097–1108.

Hameed, S. and Lebedeff, S. A. (1975). Application of integral method to heat conduction in non-homogeneous media. *ASME J. Heat Transfer* **94**, 304–305.

Hamill, T. D. and Bankoff, S. G. (1963). Maximum and minimum bounds on freezing–melting rates with time-dependent boundary conditions. *A.I.Ch.E. J.* **9**, 741–744.

Hill, J. M. (1982). *Solution of Differential Equations by Means of One-parameter Groups.* Research Notes in Mathematics, Vol. 63. Pitman, London.

Hill, J. M. (1984a). On the pseudo-steady-state approximation for moving-boundary diffusion problems. *Chem. Engng. Sci.* **39**, 187–190.

Hill, J. M. (1984b). Some unusual unsolved integral equations for moving-boundary diffusion problems. *Math. Scientist* **9**, 15–23.

Hill, J. M. and Dewynne, J. N. (1984). Improved lower bounds for the motion of moving boundaries. *J. Austral. Math. Soc.* Ser. B **26**, 165–175.

Hill, J. M. and Dewynne, J. N. (1985). A note on Langford's cylinder functions $c_n(z, z_0)$ and $e_n(z, z_0)$. *Quart. Appl. Math.* **43**, 179–185.

Hill, J. M. and Dewynne, J. N. (1986). On the inward solidification of cylinders. *Quart. Appl. Math.* **44**, 59–70.

Hill, J. M. and Kucera, A. (1983a). The time to complete reaction or solidification of a sphere. *Chem. Engng. Sci.* **38**, 1360–1362.

Hill, J. M. and Kucera, A. (1983b). Freezing a saturated liquid inside a sphere. *Int. J. Heat Mass Transfer* **26**, 1631–1637.

Hill, J. M. and Kucera, A. (1984). On boundary-fixing transformations for the classical Stefan problem. *Mech. Res. Comm.* **11**, 91–96.

Hill, J. M. and Kucera, A. (1985). An approximate solution for planar solidification with Newton cooling at the surface. *Z. angew. Math. Phy.* **36**, 637–647.

Huang, C. L. and Shih, Y. P. (1975a). Perturbation solution for planar solidification of a saturated liquid with convection at the wall. *Int. J. Heat Mass Transfer* **18**, 1481–1483.

Huang, C. L. and Shih, Y. P. (1975b). A perturbation method for spherical and cylindrical solidification. *Chem. Engng. Sci.* **30**, 897–906.

Huang, C. L. and Shih, Y. P. (1975c). Perturbation solutions of planar diffusion-controlled moving-boundary problems. *Int. J. Heat Mass Transfer* **18**, 689–695.

Hwang, C. T. (1977). On quasi-static solutions for buried pipes in permafrost. *Can. Geotech. J.* **14**, 180–191.

Kehoe, P. (1972). A moving-boundary problem with variable diffusivity. *Chem. Engng. Sci.* **27**, 1184–1185.

Kern, J. (1977). A simple and apparently safe solution to the generalized Stefan problem. *Int. J. Heat Mass Transfer* **20**, 467–474.

Kern, J. and Hansen, J. O. (1976). Transient heat conduction in cylindrical systems with an axially moving boundary. *Int. J. Heat Mass Transfer* **19**, 707–714.

Kern, J. and Wells, G. L. (1977). Simple analysis and working equations for the solidification of cylinders and spheres. *Met. Trans.* B **8**, 99–105.

Knight, J. H. and Philip, J. R. (1974). Exact solutions in non-linear diffusion. *J. Engng. Math.* **8**, 219–227.

Krishnamurthy, S. and Shah, Y. T. (1979). An approximate solution to the moving-boundary problem with two simultaneous reactions. *Chem. Engng. Sci.* **34**, 1067–1070.

Lamé, G. and Clapeyron, B. P. (1831). Mémoire sur la solidification par refroidissement d'un globe solide. *Ann. Chem. Phys.* **47**, 250–256.

Landau, H. G. (1950). Heat conduction in a melting solid. *Quart. Appl. Math.* **8**, 81–94.

Langford, D. (1966). A closed-form solution for the constant velocity solidification of spheres initially at the fusion temperature. *Brit. J. Appl. Phys.* **17**, 286.

Langford, D. (1967a). Pseudo-similarity solutions of the one-dimensional diffusion equation with applications to the phase-change problem. *Quart. Appl. Math.* **25**, 45–52.

Langford, D. (1967b). New analytic solutions of the one-dimensional heat equation for temperature and heat-flow rate both prescribed at the same fixed boundary (with applications to the phase-change problem). *Quart. Appl. Math.* **24**, 315–322.

Lardner, T. J. (1967). Approximate solutions to phase-change problems. *AIAA J.* **5**, 2079–2080.

Lardner, T. J. and Pohle, F. V. (1961). Application of the heat-balance integral to problems of cylindrical geometry. *J. Appl. Mech.* **28**, 310–312.

Lederman, J. M. and Boley, B. A. (1970). Axisymmetric melting or solidification of circular cylinders. *Int. J. Heat Mass Transfer* **13**, 413–427.

Lin, S. (1981). An exact solution of the sublimation problem in a porous medium. *J. Heat Transfer* **103**, 165–168.

Lin, S. (1982). An exact solution of the desublimation problem in a porous medium. *Int. J. Heat Mass Transfer* **25**, 625–630.

Meksyn, D. (1961). *New Methods in Laminar Boundary-layer Theory.* Pergamon Press, New York.

Mikhailov, M. D. (1975). Exact solution of temperature and moisture distributions in a porous half-space with a moving evaporation front. *Int. J. Heat Mass Transfer* **18**, 797–804.

Mikhailov, M. D. (1976). Exact solution for freezing of humid porous half-space. *Int. J. Heat Mass Transfer* **19**, 651–655.

Ockendon, J. R. (1975). Techniques of analysis. Article in Ockendon and Hodgkins (1975), 138–149.

Paschkis, V. (1953). Solidification of cylinders. *Trans. Am. Foundrymen's Soc.* **61**, 142–149.

Pedroso, R. I. and Domoto, G. A. (1973a). Exact solution by perturbation method for planar solidification of a saturated liquid with convection at the wall. *Int. J. Heat Mass Transfer* **16**, 1816–1819.

Pedroso, R. I. and Domoto, G. A. (1973b). Perturbation solutions for spherical solidification of saturated liquids. *ASME J. Heat Transfer* **95**, 42–46.

Pedroso, R. I. and Domoto, G. A. (1973c). Inward spherical solidification—solution by the method of strained coordinates. *Int. J. Heat Mass Transfer* **16**, 1037–1043.

Pekeris, C. L. and Slichter, L. B. (1939). Problem of ice formation. *J. Appl. Phys.* **10**, 135–137.

Poots, G. (1962a). An approximate treatment of a heat-conduction problem involving a two-dimensional solidification front. *Int. J. Heat Mass Transfer* **5**, 339–348.

Poots, G. (1962b). On the application of integral methods to the solution of problems involving the solidification of liquids initially at fusion temperature. *Int. J. Heat Mass Transfer* **5**, 525–531.

Prasad, A. and Agrawal, H. C. (1972). Biot's variational principle for a Stefan problem. *AIAA J.* **10**, 325–327.

Prasad, A. and Agrawal, H. C. (1974). Biot's variational principle for aerodynamic ablation of melting solids. *AIAA J.* **12**, 250–252.

Riley, D. S. and Duck, P. W. (1977). Application of the heat-balance integral method to the freezing of a cuboid. *Int. J. Heat Mass Transfer* **20**, 294–296.

Riley, D. S., Smith, F. T. and Poots, G. (1974). The inward solidification of spheres and circular cylinders. *Int. J. Heat Mass Transfer* **17**, 1507–1516.

Savino, J. M. and Siegel, R. (1969). An analytical solution for solidification of a moving warm liquid onto an isothermal cold wall. *Int. J. Heat Mass Transfer* **12**, 803–809.

Selim, M. S. and Seagrave, R. C. (1973a). Solution of moving-boundary transport problems in finite media by integral transforms. I. Problems with a plane moving boundary. *Ind. Engng. Chem. Fundam.* **12**, 1–8.

Selim, M. S. and Seagrave, R. C. (1973b). Solution of moving-boundary transport problems in finite media by integral transforms. II. Problems with a cylindrical or spherical moving boundary. *Ind. Engng. Chem. Fundam.* **12**, 9–13.

Selim, M. S. and Seagrave, R. C. (1973c). Solution of moving-boundary transport problems in finite media by integral transforms. III. The elution kinetics of the copper ammine complex from a cation-exchange resin. *Ind. Engng. Chem. Fundam.* **12,** 14–17.

Shanks, D. (1955). Non-linear transformations of divergent and slowly convergent sequences. *J. Math. Phys.* **34,** 1–42.

Shaw, R. P. (1974). An integral equation approach to diffusion. *Int. J. Heat Mass Transfer* **17,** 693–699.

Shih, Y. P. and Chou, T. C. (1971). Analytical solutions for freezing a saturated liquid inside or outside spheres. *Chem. Engng. Sci.* **26,** 1787–1793.

Shih, Y. P. and Tsay, S. Y. (1971). Analytical solutions for freezing a saturated liquid inside or outside cylinders. *Chem. Engng. Sci.* **26,** 809–816.

Siegel, R. (1968). Conformal mapping for steady two-dimensional solidification on a cold surface in flowing liquid. U.S. National Aeronautics and Space Administration, Technical Note D-4771, 1–19.

Siegel, R. (1978). Analysis of solidification interface shape during continuous casting of a slab. *Int. J. Heat Mass Transfer* **21,** 1421–1430.

Siegel, R. and Savino, J. M. (1966). An analysis of the transient solidification of a flowing warm liquid on a convectively cooled wall. *Proc. ASME 3rd Int. Heat Transfer Conference,* Vol. 4, 141–151.

Solomon, A. D. (1978). The applicability and extendability of Megerlin's method for solving parabolic free-boundary problems. Article in Wilson *et al.* (1978), 187–202.

Solomon, A. D. (1979). A relation between surface temperature and time for a phase-change process with a convective boundary condition. *Letters Heat Mass Transfer* **6,** 189–197.

Solomon, A. D. (1980). On the melting time of a simple body with a convection boundary condition. *Letters Heat Mass Transfer* **7,** 183–188.

Solomon, A. D. (1981a). A note on the Stefan number in slab melting and solidification. *Letters Heat Mass Transfer* **8,** 229–235.

Solomon, A. D. (1981b). On surface effects in heat transfer calculations. *Computers Chem. Engng.* **5,** 1–5.

Soward, A. M. (1980). A unified approach to Stefan's problem for spheres and cylinders. *Proc. R. Soc. Lond.* A **373,** 131–147.

Sparrow, E. M. (1960). The melting of finite slabs. *ASME Trans.* **82,** 598.

Sproston, J. L. (1981). Two-dimensional solidification in pipes of rectangular section. *Int. J. Heat Mass Transfer* **24,** 1493–1501.

Stefan, J. (1889a). Über einige Probleme der Theorie der Wärmeleitung. *S.-B. Wien. Akad. Mat. Natur.* **98,** 473–484.

Stefan, J. (1889b). Über die Diffusion von Säuren und Basen gegen einander. *S.-B. Wein. Akad. Mat. Natur.* **98,** 616–634.

Stefan, J. (1889c). Über die Theorie der Eisbildung, insbesondere über die Eisbildung im Polarmeere. *S.-B. Wien. Akad. Mat. Natur.* **98,** 965–983.

Stefan, J. (1889d). Über die Verdampfund und die Auflösung als Vorgänge der Diffusion. *S.-B. Wien. Akad. Mat. Natur.* **98,** 1418–1442.

Stefan, J. (1891). Über die Theorie der Eisbildung, insbesondere über die Eisbildung im Polarmeere. *Annalen der Physik und Chemie* **42,** 269–286.

Stewartson, K. and Waechter, R. T. (1976). On Stefan's problem for spheres. *Proc. R. Soc. Lond.* A **348**, 415–426.

Tait, R. J. (1979). Additional pseudo-similarity solutions of the heat equation in the presence of moving boundaries. *Quart. Appl. Math.* **37**, 313–324.

Tao, L. C. (1967). Generalized numerical solutions of freezing a saturated liquid in cylinders and spheres. *A.I.Ch.E. J.* **13**, 165–169.

Tao, L. N. (1978). The Stefan problem with arbitrary initial and boundary conditions. *Quart. Appl. Math.* **36**, 223–233.

Tao, L. N. (1979). On solidification problems including the density jump at the moving boundary. *Q. J. Mech. Appl. Math.* **32**, 175–185.

Tao, L. N. (1980). On solidification of a binary alloy. *Q. J. Mech. Appl. Math.* **33**, 211–225.

Tao, L. N. (1981). The exact solutions of some Stefan problems with prescribed heat flux. *J. Appl. Mech.* **48**, 732–736.

Tao, L. N. (1982a). The Stefan problem with an imperfect thermal contact at the interface. *J. Appl. Mech.* **49**, 715–720.

Tao, L. N. (1982b). The Cauchy–Stefan problem. *Acta Mechanica* **45**, 49–64.

Tayler, A. B. (1975). The mathematical formulation of Stefan problems. Article in Ockendon and Hodgkin (1975), 120–137.

Tayler, A. B. (1982). Differential equations and the real world. *Bull. Austral. Math. Soc.* **26**, 421–443.

Theofanous, T. G. and Lim, H. C. (1971). An approximate analytical solution for non-planar moving-boundary problems. *Chem. Engng. Sci.* **26**, 1297–1300.

Voller, V. R. and Cross, M. (1981). Estimating the solidification/melting times of cylindrically symmetric regions. *Int. J. Heat Mass Transfer* **24**, 1457–1462.

Weber, H. (1912). *Die partiellen Differentialgleichungen der mathematischen Physik.* 5th Ed., Vol. 2. Friedr. Vieweg & Sohn, Braunschweig, Germany.

Weinbaum, S. and Jiji, L. M. (1977). Singular perturbation theory for melting or freezing in finite domains initially not at the fusion temperature. *J. Appl. Mech.* **44**, 25–30.

Wilson, D. G. (1982). Lagrangian coordinates for moving-boundary problems. *SIAM J. Appl. Math.* **42**, 1195–1201.

Wu, T. S. (1966). Bounds in melting slabs with several transformation temperatures. *Q. J. Mech. Appl. Math.* **19**, 183–195.

Yagoda, H. P. and Boley, B. A. (1970). Starting solutions for melting of a slab under plane or axisymmetric hot spots. *Q. J. Mech. Appl. Math.* **23**, 225–246.

Yan, M. M. and Huang, P. N. S. (1979). Pertubation solutions to phase-change problem subject to convection and radiation. *J. Heat Transfer* **101**, 96–100.

Yeh, L. T. and Chung, B. T. F. (1977). Phase change in a radiating and convecting medium with variable thermal properties. *J. Spacecraft and Rockets* **14**, 178–182.

Yeh, L. T. and Chung, B. T. F. (1979). A variational analysis of freezing or melting in a finite medium subject to radiation and convection. *ASME J. Heat Transfer* **101**, 592–597.

Yuen, W. W. (1980). Application of the heat-balance integral to melting problems with initial subcooling. *Int. J. Heat Mass Transfer* **23**, 1157–1160.

Zener, C. (1949). Theory of growth of spherical precipitates from solid solution. *J. Appl. Phys.* **20**, 950–953.

Zyszkowski, W. (1969). The transient temperature distribution in one-dimensional heat-conduction problems with non-linear boundary conditions. *ASME J. Heat Transfer* **91**, 77–82.

Citation index